Elementary Statistics

Elementary Statistics

B. C. ERRICKER, B.Sc

Formerly Head of the Mathematics Department, High School, Leyton

Revised by

W. T. ELLIS, B.Sc

Formerly Head of the Management and Business Studies Department, Bath Technical College

and

PAT DAVIES, B.Sc(Econ.)

Lecturer in Business Studies, Bath Technical College

HODDER AND STOUGHTON

LONDON   SYDNEY   AUCKLAND   TORONTO

Erricker, Brinley Cenydd
  Elementary statistics. – 3rd ed.
  1. Mathematical statistics
  I. Title   II. Ellis, W. T.   III. Davies, Pat
  519.5   QA276
  ISBN 0 340 25683 4

First printed 1965
Second edition 1970
Reprinted 1973, 1974, 1975, 1976, 1977, 1978 (twice)
Third edition 1981
Reprinted 1982

Printed in Great Britain for
Hodder and Stoughton Educational,
a division of Hodder and Stoughton Ltd.,
Mill Road, Dunton Green, Sevenoaks, Kent TN13 2YD
by the University Press, Cambridge

Preface

This book is an elementary introduction to the study of statistics. It is designed:
  (i) to meet the requirements of students preparing for statistics as an Ordinary Level subject for the General Certificate of Education,
 (ii) as a useful course of study for anyone requiring an understanding of statistics as a branch of some other study, and whose knowledge of mathematics does not reach the heights necessary to read the standard works on the subject,
(iii) for those teachers who realise that pupils today require a knowledge of the elementary principles and techniques of statistics as a necessary part of a well-balanced education, and have felt the need of a text-book requiring only an elementary knowledge of mathematics for its study, and with ready-made examples and exercises graded to meet the requirements of a developing pupil.

It is my experience that many pupils prefer the study of this 'practical' branch of mathematics to the more abstract parts, and I am very grateful to the very many pupils who have been 'guinea pigs' for the testing of various forms of presentation. I, and I believe they, have derived much pleasure from the experiment. Statistics should be a 'practical' subject, and many of the examples in the book should be supplemented by practical exercises on data collected by the student so that he can learn to classify, tabulate and analyse 'raw' material. Sometimes this involves much arithmetical computation, but I find that interest in the exercise makes pupils very tolerant of this work.

My chief qualification for writing this book is that I have had to worry through most of the difficulties of presentation alone, and have become aware of the pupils' difficulties. Under the circumstances I realise that it has been impossible to write upon the subject without unwittingly incorporating parts of the work of other authors, and I wish to record the benefit and pleasure I have received from them, and in particular Yule and Kendall's *Theory of Statistics*.

For permission to use questions set in examinations I wish to thank those organisations listed on p. 8, and I also wish to thank the Controller of H.M. Stationery Office for permission to publish material from the Annual Abstract of Statistics, the Monthly Digest of Statistics, and Economic Trends. All such material has been denoted by the symbol (A).

Care has been taken to avoid mistakes, but if any are found in the text or answers the fault is mine, and I should be pleased if they are brought to my notice. In some cases permission to reproduce answers to the questions in this book has not been granted by the relevant examining board, and these answers have been omitted.

Preface to the Third Edition

One of the main purposes of this edition has been to present up-to-date figures in questions, exercises, charts, tables and diagrams. In practically every case the new data used for this have been taken from officially published statistics as opposed to hypothetical values, in the hope that the new material will appear the more readily relevant to the reader. Acknowledgement of the source of the data is made in each case.

The text of the previous editions has been unaltered in any essential detail but where it was thought necessary to clarify or add to an explanation, this has been included. Two examples of this are seen in the sections on 'Properties of a Sample' and 'Standardised Examination Marks'.

To meet the needs of some syllabuses an additional section dealing with Probability Trees has been introduced in the chapter on the Binomial Distribution. Every effort has been made to identify and correct any computational error that may have existed in the second edition, particularly in the answers to the exercises. We, however, know intuitively that the probability that undetected errors still exist is greater than zero and we would be grateful to the reader who finds evidence to confirm that our intuition is correct.

Finally, we wish to thank the various examining bodies for permission to use questions from past examination papers and to produce answers. We also thank the Controller of H.M. Stationery Office for permission to use material from the Annual Abstract and the Monthly Digest of Statistics.

W.T.E.
P.S.D.

Contents

Note: Chapter 20 may be read before Chapter 19

The following abbreviations indicate the source of the questions set in examinations:

L	The University of London School Examinations Department, Ordinary Level.
LA.	The University of London School Examinations Department, Advanced Level.
C.	The University of Cambridge Local Examinations Syndicate, Ordinary Level.
C.A.	The University of Cambridge Local Examinations Syndicate, Advanced Level.
A.E.B.	The Associated Examining Board for the General Certificate of Education.
J.M.B.	The Northern Universities Joint Matriculation Board.
R.S.A.	The Royal Society of Arts Examination Department, Stage I Elementary.
R.S.A.I.	The Royal Society of Arts Examination Department, Intermediate Level.
R.S.A.A.	The Royal Society of Arts Examination Department, Advanced Level.
I.T.	The Institute of Transport.
P.M.	The Institute of Personnel Management.
I.S.A.	The Institute of Statisticians, Part I Examination.
I.S.I.	The Institute of Statisticians, Part II Examination.
L.G.I.	The Local Government Training Board, Intermediate Level.
L.G.F.	The Local Government Training Board, Final Level.

1 The Development and Meaning of Statistics

Statistics and the State
The word 'statistics' is derived from the word 'statist' which has from early times been used to describe a person skilled in State affairs. Governments have always found it necessary to collect facts relating to State affairs, and modern governments find it essential to have a reliable knowledge of all that goes on within the State. This has led to the greater use of number in describing matters of State. The method of collection and the analysis of these numbers have been developed into a branch of knowledge known as 'statistics'.

Development of Statistics
As far as is known the word 'statistics' was first used by a German writer about 1770, and was defined as 'the Science that teaches us what is the political arrangement of all the modern States of the new world'. The facts were given in a *verbal* form, and it was not until the early nineteenth century that *numerical* facts began to be collected. Statistics was then defined as 'the setting forth of the characteristics of a State by numerical methods'.

In 1834 the Royal Statistical Society was founded, and in its first journal, issued in 1838–9, statistics was defined as 'the ascertaining and bringing together of those facts which are calculated to illustrate the condition and prospects of *society*', and that 'the Statist commonly prefers to *employ figures and tabular exhibitions*'. Thus statistics which started as verbal descriptions was now mostly concerned with numerical facts.

Further changes of meaning took place and the word came to stand for the numbers themselves. Thus figures relating to society are known as 'sociological statistics', to education as 'educational statistics', and so on.

Lack of Control
The great difficulty about collecting facts relating to subjects similar to those mentioned in the last paragraph is the lack of control over the subjects. The scientist when performing an experiment can generally arrange matters so that he can examine one fact at a time. For instance,

when investigating the effect of pressure on a gas he finds it a simple matter to keep the mass and temperature of the gas constant and measure the change in volume, but the statistician cannot do this. For instance, if investigating facts relating to population he will find it impossible to prevent the occurrence of either births, deaths, marriages, movements within the population, emigration, or immigration during the period of the investigation, and thus his facts can only be approximate —*his numbers are affected by uncontrolled variation.*

The mathematical approach to this type of fact is of recent development. There is no definition that includes all that is meant by modern statistics, but a suitable working definition is 'the classification, tabulation, and analysis of numerical facts affected by many factors'.

Terms

It is important to understand the following terms, and the distinction between their meanings:

Statistic.   An estimate from a set of observations, e.g., a cricketer's average score, the average age of a form, the daily average number of pupils in a school.

Statistics.   This has two meanings: (1) the plural of statistic, that is numerical facts, generally described by the subject to which they apply, e.g. educational statistics, labour statistics, biological statistics; (2) the subject that deals with the classification, tabulation, and analysis of numerical facts affected by many factors.

Warning

The methods used to make clear the facts of statistics are similar to the methods used in mathematics. A word of warning must be given. The successes obtained by the application of mathematics to science have given many people the impression that the application of mathematical methods to any set of numbers produces a result of extreme accuracy. This is not so. The results cannot be more accurate than the figures from which they are obtained, and an honest person never gives his results to a degree of accuracy not justified by the original figures. It is one of the purposes of statistics to estimate the accuracy of, and the reliance that can be placed on, any statistic.

2 Accuracy

Significant Figures

No measurement is ever exact. If the length of a page of a book is measured and the answer given as 21·6 cm then, as far as we know, the length lies between 21·55 cm and 21·65 cm. The fourth figure cannot be guaranteed and should not be given. Measurements should only be given 'correct as far as they go', and in this case the answer should be given as 21·6 cm. The answer is then said to be correct to three *significant figures*.

Thus 492 cm, 4·92 metres, and 0·00492 kilometres are results quoted to three significant figures, and are of the same degree of accuracy. The noughts at the beginning of 0·00492 kilometres are due to the change in units, and are *not significant*.

Example   Express correct to two significant figures
1.   3·1681   2.   0·00031681

Answer
1.   The first two figures are 3·1, but 3·1681 is nearer to 3·2 than 3·1.
     Therefore 3·1681 = 3·2 to two significant figures.
2.   As the first noughts are not significant the answer is 0·00032.

Example   Express the following correct to one significant figure:
1.   2·73   2.   28·4   3.   0·76   4.   921   5.   976   6.   0·98

Answer
1.   3   2.   30   3.   0·8   4.   900   5.   1000   6.   1

Exercise 2.1

Express the following correct to the number of significant figures given in italics:

1. 4·762 *3*	2. 31·36 *2*	3. 0·851 *2*
4. 5763 *3*	5. 6021 *2*	6. 763 *1*
7. 9784 *1*	8. 0·00374 *1*	9. 0·00402 *2*
10. 200·62 *3*	11. 200·12 *3*	12. 999·6 *3*

11

Noughts at the End of a Number

If a measurement is given as 170 cm, it is impossible to know the degree of accuracy intended. If the measurement is given correct to two significant figures the measurement lies between 165 cm and 175 cm, and the final nought is not significant, but if the answer is correct to three significant figures the measurement lies between 169·5 cm and 170·5 cm, and the final nought is significant.

Thus noughts at the end of a number sometimes offer difficulty, but it is sometimes possible to decide whether they are significant or not from the way in which the number is presented.

Thus 'the issued capital of a company is £3000', suggests £3000 to be an exact number, and the noughts are significant, but 'the amount of coal in reserve is 3000 tonnes', is probably only significant to one figure. A measurement given as 1·70 should not be stated as 1·7, as 1·70 is ten times more accurate than 1·7.

Exercise 2.2

State to how many significant figures the numbers in the following statements are probably quoted:

1. one pound equals 100 pennies;
2. a ship's displacement is 3000 tonnes;
3. the manager's salary is £9500 per year;
4. the manager's expenses amount to close on £500 a year;
5. the population of Great Britain is 50000000;
6. there are 50 matches in a box;
7. there are 20 cigarettes in a packet.

Alternative Method

Another method of indicating the accuracy of a number is to quote the limits of possible error.

Example   A boy's height is 163 cm correct to the
nearest cm. Indicate the possible error.

· The boy's height lies between 162·5 cm, and 163·5 cm, and may be written as 163 ± 0.5 cm (read as 163 plus or minus 0·5 cm).

Exercise 2.3

Indicate the possible limits of error in the following:
1. the length of a table is 2·73 metres correct to the nearest centimetre;
2. the height of a building is 12 m correct to the nearest metre;
3. a runner's time for 100 m is 10·2 sec, correct to $\frac{1}{10}$th second;

4. the number of pages in a book is 532;
5. the number of pupils on the school register is 693.

The Effect of Addition on Accuracy
Example
The sides of a rectangle are 1·6 and 6·7 cm long. Find their sum.

The sides are given to two significant figures. Therefore their lengths lie between 1·55 and 1·65 cm, and 6·65 and 6·75 cm, or $1·6 \pm 0·05$ and $6·7 \pm 0·05$ cm.

Therefore their sum is $(1·6 \pm 0·05) + (6·7 \pm 0·05) = 8·3 \pm 0·1$ cm.

All we can say is that the sum lies between 8·2 cm and 8·4 cm, and is therefore only correct to one significant figure.

The Effect of Subtraction on Accuracy
Example
Find the difference in the lengths of the sides of the above rectangle.

The biggest possible difference is between 6·75 and 1·55 cm, i.e. 5·20 cm, and the smallest possible difference is between 6·65 and 1·65 cm, i.e. 5·00 cm.

Therefore all we can say is the difference lies between 5·00 and 5·20 cm, or is $5·1 \pm 0·1$ cm. The result is only correct to one significant figure.

The Effect of Multiplication on Accuracy
Example
Find the area of the rectangle.

The largest possible area is $6·75 \times 1·65$ cm², i.e. 11·1375 cm², and the smallest possible area is $6·65 \times 1·55$ cm², i.e. 10·3075 cm². The mean of these two results is 10·7225 cm². So the result can be given as $10·7225 \pm 0·415$ cm², or with sufficient accuracy as $10·7 \pm 0·4$ cm². The area is correct to one significant figure.

The Effect of Division on Accuracy
Example
What is the result of dividing the length of the longer side by the length of the shorter side?

The largest answer is $(6·75)/(1·55)$, or 4·35 to two decimal places, and the smallest answer is $(6·65)/(1·65)$ or 4·03 to two decimal places. Therefore the answer is correct only to one significant figure. It could be written as $4·2 \pm 0·2$.

Example

The following numbers are given to two significant figures. Find their sum in the form $N \pm x$.

$$3100 + 270 + 43.$$

$$\text{Sum} = (3100 \pm 50) + (270 \pm 5) + (43 \pm 0 \cdot 5)$$
$$= 3413 \pm 55 \cdot 5.$$

Exercise 2.4

In the following, 1 to 13, the numbers are given in significant figures. Find the value of each in the form $N \pm x$.

1. $31 \cdot 6 + 29 \cdot 1$
2. $5 \cdot 32 + 1 \cdot 59$
3. $91 \cdot 4 + 73 \cdot 2$
4. $1 \cdot 29 + 0 \cdot 73$
5. $3 \cdot 51 + 42 \cdot 6 + 21$
6. $53 \cdot 47 - 21 \cdot 36$
7. $26 - 7 \cdot 3$
8. $11 \cdot 1 + 7 \cdot 2 - 3 \cdot 6 - 4 \cdot 7$
9. $9 \cdot 72 \times 4 \cdot 12$
10. $0 \cdot 0472 \times 0 \cdot 42$
11. $9 \cdot 73 \div 2 \cdot 45$
12. $50 \cdot 1 \div 5 \cdot 0$
13. $(21 \cdot 2 \times 36 \cdot 3)/10 \cdot 1$

14. Between what limits does the sum of £0·42, £0·36, £0·10, £0·93 lie if each amount is correct to the nearest penny?

15. 4·7 cm, 6·3 cm, 4·1 cm, 5·2 cm, are the lengths of the sides of a quadrilateral correct to the nearest mm. Between what limit does the length of its perimeter lie?

16. The volume of a pyramid is one-third the area of its base multiplied by its height. If the base is a square of side 5·2 cm, and the height is 6·3 cm (both measurements given to the nearest mm), find between what limits the volume lies.

17. The length of a running track is 100 ± 1 m, and the time taken by a sprinter to run along it is 10·2 sec correct to the nearest 0·1 sec. Between what limits does the sprinter's average speed in m/s^{-1} lie?

18. The death rate of the population of a certain town was stated to be 14·5 per thousand when the population was known to be 21 000 expressed to the nearest thousand. Calculate the number of deaths during the year and show the degree of accuracy in your answer. *(L)*

19. State between what limits the following product lies if the numbers are rounded off to the given significant figures:

$$9 \cdot 73 \times 2 \cdot 41. \qquad (A.E.B.)$$

20. It is known that 5·3 (correct to the first decimal place) per thousand of the articles produced by a firm are faulty and have to be rejected. In 1963 the firm produced 65 000 articles (correct

to the nearest 1000). Calculate the limits between which the actual number of articles fit for sale lies. (*A.E.B.*)

21. The crude death rate for a town was stated to be 13·2 per thousand, the rate being rounded off to one decimal place, and the population of the town, to the nearest one thousand, was 35000. Calculate the limits between which the actual number of deaths lay.

(*A.E.B.*)

22. State between what limits the following product lies if the numbers are rounded off to the given significant figures. Give your answers correct to the nearest integer

$$3·7 \times 2·6 \times 4·2.$$ (*A.E.B.*)

23. State between what limits (correct to two decimal places) the quotient of the following division lies if the numbers are rounded off to the given significant figures:

$$9·73 \div 2·45.$$ (*A.E.B.*)

24. A baker stores flour in 50 kg sacks, and the mass of flour in a sack is correct to within 1 per cent. The flour is measured into $\frac{1}{2}$ kg bags correct to the nearest 1/100 kg. Calculate
    (i) the smallest number
    (ii) the largest number
of bags the baker could fill from a 50 kg sack. (*A.E.B.*)

25. The populations of England and Wales, Scotland, and Northern Ireland are given to the degree of accuracy stated in Table 1. Between what limits does the population of the United Kingdom lie? Give your answer in the form $N \pm x$.

Table 1

Country	Population	Degree of accuracy
England and Wales	46166391	Correct to within 5%
Scotland	5184000	Correct to nearest 1000
Northern Ireland	1428421	Correct to within 3%

(*A.E.B.*)

26. A farmer sows 254 hectares (correct to the nearest hectare) of wheat and produces 4·08 m³ (correct to the nearest 0·01 m³) of wheat per hectare. One m³ of wheat has a mass of 1·5 kg (correct to the nearest 0·1 kg).

    Find the limits, correct to the nearest 10 kg, between which his production lies. (*A.E.B.*)

27. The population of a town in 1968 was 135000. The population increase between 1968 and 1978 was 6 per cent of the 1968 figure. Estimate the upper and lower limits of the population in 1978 if

the population figure is given correct to the nearest 1000 and the percentage increase to the nearest one per cent.

If the population increase between 1958 and 1968 was 5 per cent (correct to the nearest one per cent) of the 1958 value, estimate the upper and lower limits of the population in 1958. (*A.E.B.*)

28. Find the limits of the following decimal fraction if the numbers are rounded off to the first decimal place.

$$\frac{3\cdot7 \times 1\cdot2}{2\cdot1}$$

(*A.E.B.*)

Rounding off Numbers

So far we have considered the maximum error that can be introduced in a simple numerical problem. Sometimes when dealing with numbers it is found convenient to round the numbers off to the nearest ten, or hundred, or thousand, and so on. In cases like this the maximum error seldom occurs, and, generally, the larger the number of items the smaller the error.

It is highly desirable that a constant procedure is adopted in rounding off numbers. The method is best illustrated by an example.

Example

Round off the numbers 341, 498, 365, 135, 608, to the nearest ten.

341, 498, 608 offer no difficulty. They become 340, 500, 610.

The numbers 365, 135 offer a choice between 360, 130, and 370, 140. In such cases it is customary to round off the numbers so that the last number is even; that is, 365 becomes 360, and 135 becomes 140.

Exercise 2.5

From a list of Premium Bond winning numbers write down 10 numbers consisting of the last three figures of the winning numbers, and find their sum.

Round off the numbers to the nearest ten and find their sum. Calculate the percentage error introduced by rounding off the numbers.

Repeat the above process ten times, each time selecting a different group of ten numbers.

Repeat the process for the hundred numbers.

Round off the numbers to the nearest hundred and find the percentage error in the sum.

What are your conclusions?

Relative and Percentage Errors

The errors considered so far are called absolute errors. Sometimes errors are expressed as 'relative errors' and 'percentage errors'. If e is the absolute error in an estimated value x, then $\frac{e}{x}$ is the relative error, and $\frac{e}{x} \times 100$ is the percentage error.

Thus in the example on page 14, the magnitude of the absolute error in the estimate 3413 is 55·5. Therefore, the magnitude of the relative error is $\frac{55·5}{3413}$, which is nearly equal to 0·0163, and the magnitude of the percentage error nearly equals 1·63.

An absolute error does not give the same information as a relative or percentage error. Thus, if the population of one village is given as 110 instead of 100, the absolute error is 10 and the relative and percentage errors are 1/10 and 10%. If the population of another village is given as 1050 instead of 1000, the absolute error is five times as great, that is 50, but the relative and percentage errors are one half the previous relative and percentage errors, that is, $\frac{50}{1000} = \frac{1}{20}$ and 5%, respectively.

Approximations and Calculating Errors

If the errors are small in comparison with the estimated values of the quantities, approximations can be made in the calculations. The assumption made is that the error in any of the quantities is so small that squares of their relative values may be neglected.

Let us first consider the approximate values of $\frac{1}{1+a}$ and $\frac{1}{1-a}$ where a is small in comparison with 1. If we divide 1 by $1+a$ we have

$$1+a)1 \qquad (1-a+a^2+\text{higher powers of } a.$$
$$\underline{1+a}$$
$$-a$$
$$\underline{-a-a^2}$$
$$a^2$$

Thus, to a first approximation, that is neglecting values of a above the first power, we have

$$\frac{1}{1+a} \simeq 1-a.$$

Similarly, to a first approximation

$$\frac{1}{1-a} \simeq 1+a.$$

Writing more simply we have

$$\frac{1}{1 \pm a} \simeq 1 \mp a$$

meaning that the plus sign on the left-hand side of the equals sign and the minus sign on the right-hand side of the equals sign must go together, and vice versa. If $a = 0.01$ then $\dfrac{1}{1+0.01} = 0.9900990099\ldots$, and if we use the approximate method we get $1 - 0.01 \simeq 0.99$ and neglect 0.000099. Similarly $\dfrac{1}{1-0.001} = 1.0101\ldots$, and approximately $\simeq 1.01$.

Remember that to a first approximation

$$\frac{1}{1 \pm a} \simeq 1 \mp a.$$

Addition and Subtraction

Suppose e_1 and e_2 are the absolute errors in the estimated values of two quantities x_1 and x_2, respectively. Then the greatest absolute value of the sum of the quantities is

$$(x_1 + e_1) + (x_2 + e_2) = (x_1 + x_2) + (e_1 + e_2)$$

and the smallest absolute value is

$$(x_1 - e_1) + (x_2 - e_2) = (x_1 + x_2) - (e_1 + e_2)$$

The greatest value of the difference between the quantities is

$$(x_1 + e_1) - (x_2 - e_2) = (x_1 - x_2) + (e_1 + e_2)$$

and the smallest value of the difference between the quantities is

$$(x_1 - e_1) - (x_2 + e_2) = (x_1 - x_2) - (e_1 + e_2)$$

Therefore, the largest absolute error in both cases is

$$e_1 + e_2$$

By definition the relative error for the sum is $\dfrac{e_1 + e_2}{x_1 + x_2}$, and for the difference $\dfrac{e_1 + e_2}{x_1 - x_2}$, and the corresponding percentage errors are $\dfrac{e_1 + e_2}{x_1 + x_2} \times 100$ and $\dfrac{e_1 + e_2}{x_1 - x_2} \times 100$.

Example

The absolute errors in the estimates of 500 and 200 are 5 and 1, respectively. Calculate the largest error in (i) their sum, (ii) their difference.

(i)

The largest absolute error in the sum $= 5 + 1 = 6$.

The largest relative error in the sum $= \dfrac{6}{500 + 200} = 0.0086$.

The largest percentage error in the sum $\quad = 0\cdot86.$
(ii)
The largest absolute error in the difference $\quad = 5+1 = 6$

The largest relative error in the difference
$$= \frac{6}{500-200}$$
$$= 0\cdot02$$

The largest percentage error in the difference $= 2$

Multiplication
If e_1 and e_2 are small in comparison with x_1 and x_2, then $\dfrac{e_1}{x_1}$ and $\dfrac{e_2}{x_2}$ are
small quantities such that their squares and products may be neglected.
The largest value of the product of x_1 and x_2 is
$$(x_1+e_1)(x_2+e_2) = x_1 x_2 \left(1+\frac{e_1}{x_1}\right)\left(1+\frac{e_2}{x_2}\right)$$
$$= x_1 x_2 \left(1+\frac{e_1}{x_1}+\frac{e_2}{x_2}+\frac{e_1 e_2}{x_1 x_2}\right)$$

Now $\dfrac{e_1 e_2}{x_1 x_2}$ is of the second order and if we approximate to the first

order $\dfrac{e_1 e_2}{x_1 x_2}$ can be neglected and we can assume the product is

$$x_1 x_2 \left(1+\frac{e_1}{x_1}+\frac{e_2}{x_2}\right)$$

Similarly the smallest value of the product is
$$x_1 x_2 \left[1-\left(\frac{e_1}{x_1}+\frac{e_2}{x_2}\right)\right]$$

Therefore, the absolute error of the product is approximately
$$x_1 x_2 \left(\frac{e_1}{x_1}+\frac{e_2}{x_2}\right)$$
$$= x_2 e_1 + x_1 e_2.$$

Example
If $12\cdot3$ and $4\cdot25$ are given in significant figures, what is the magnitude, correct to a first approximation, of the error in their product?

The absolute error in $12\cdot3$ is $0\cdot05$ and in $4\cdot25$ it is $0\cdot005$. Therefore the approximate absolute error in the product is
$$12\cdot3 \times 0\cdot005 + 4\cdot25 \times 0\cdot05 = 0\cdot2740$$

The approximate relative error is $\dfrac{0\cdot2740}{12\cdot3 \times 4\cdot25} = 0\cdot0052.$

The approximate percentage error is $0\cdot52\%$.

The calculation of a product can be simplified, thus,

the approximate error in the product is $x_1 x_2 \left(\dfrac{e_1}{x_1} + \dfrac{e_2}{x_2} \right)$,

the approximate relative error in the product is

$$\frac{x_1 x_2 \left(\dfrac{e_1}{x_1} + \dfrac{e_2}{x_2} \right)}{x_1 x_2} = \left(\frac{e_1}{x_1} + \frac{e_2}{x_2} \right)$$

that is, the sum of the relative errors in x_1 and x_2, and the approximate

percentage error is $\dfrac{e_1}{x_1} \times 100 + \dfrac{e_2}{x_2} \times 100$, i.e. the sum of the percentage

errors in x_1 and x_2.

Examples

The relative errors in the quantities of the last example are $\dfrac{0\cdot05}{12\cdot3}$ and

$\dfrac{0\cdot005}{4\cdot25}$, that is $0\cdot004065$ and $0\cdot0011765$. Therefore, the approximate

relative error in the product is the sum of these, i.e. $0\cdot0052$. Similarly the approximate percentage erro is the sum of $0\cdot4065$ and $0\cdot11765$, i.e. $0\cdot52\%$.

Division

The largest value of the quotient of $\dfrac{x_1}{x_2}$ is $\dfrac{x_1 + e_1}{x_2 - e_2}$, and the smallest value

$\dfrac{x_1 - e_1}{x_2 + e_2}$.

Now, $\dfrac{x_1 + e_1}{x_2 - e_2} = \dfrac{x_1 (1 + e_1/x_1)}{x_2 (1 - e_2/x_2)}$.

Since e_2 is small compared with x_2, e_2/x_2 is a small quantity, and to a

first approximation $\dfrac{1}{1 - e_2/x_2} \simeq 1 + e_2/x_2$.

Therefore, $\dfrac{x_1 (1 + e_1/x_1)}{x_2 (1 - e_2/x_2)} \simeq \dfrac{x_1}{x_2} (1 + e_1/x_1)(1 + e_2/x_2)$

$$\simeq \frac{x_1}{x_2} \left(1 + \frac{e_1}{x_1} + \frac{e_2}{x_2} \right) \quad \text{neglecting} \quad \frac{e_1 e_2}{x_1 x_2}$$

and the approximate absolute error is

$$\frac{x_1}{x_2} \left(1 + \frac{e_1}{x_1} + \frac{e_2}{x_2} \right) - \frac{x_1}{x_2} = \frac{x_1}{x_2} \left(\frac{e_1}{x_1} + \frac{e_2}{x_2} \right)$$

$$= \frac{e_1}{x_2} + \frac{x_1 e_2}{x_2^2}$$

$$= \frac{e_1 x_2 + x_1 e_2}{x_2^2}$$

The approximate relative error is

$$\frac{x_1}{x_2}\left(\frac{e_1}{x_1}+\frac{e_2}{x_2}\right)\div\frac{x_1}{x_2}=\frac{e_1}{x_1}+\frac{e_2}{x_2}$$

that is, the sum of the relative errors in x_1 and x_2.

The approximate percentage error is $\left(\dfrac{e_1}{x_1}+\dfrac{e_2}{x_2}\right)\times 100$, that is the sum of

the percentage errors in x_1 and x_2.

Similarly with the second quotient the approximate errors are the same.

Example

Calculate to a first approximation the error in $\dfrac{12\cdot 3}{4\cdot 25}$ if the numbers are

given in significant figures.

The greatest error in $12\cdot 3$ is $0\cdot 05$ and in $4\cdot 25$ is $0\cdot 005$.

The absolute error is $\dfrac{12\cdot 3\times 0\cdot 005+4\cdot 25\times 0\cdot 05}{(4\cdot 25)^2}=0\cdot 015$, the relative

error = absolute error $\times x_2\div x_1 \simeq \dfrac{0\cdot 015\times 4\cdot 25}{12\cdot 3}\simeq 0\cdot 0052$, and

percentage error $\simeq 0\cdot 52$.

An alternative solution is found as follows:

the relative errors in the numbers are $0\cdot 004\,065$ and $0\cdot 001\,176\,5$. Therefore, the relative error in the quotient is

$$0\cdot 004\,065+0\cdot 001\,176\,5\simeq 0\cdot 0052$$

and of the percentage error

$$0\cdot 4065+0\cdot 11765\simeq 0\cdot 52.$$

TABLE OF RESULTS

e_1 and e_2 are the absolute errors in x_1 and x_2 respectively,

r_1 and r_2 are the relative errors in x_1 and x_2 respectively, that is

$$r_1=\frac{e_1}{x_1} \text{ and } r_2=\frac{e_2}{x_2}$$

p_1 and p_2 are the percentage errors in x_1 and x_2 respectively, that is

$$p_1=\frac{e_1}{x_2}\times 100 \text{ and } p_2=\frac{e_2}{x_2}\times 100$$

Magnitude of Largest Error

Operation	Absolute	Relative	Percentage	Accuracy
x_1+x_2	e_1+e_2	$\dfrac{e_1+e_2}{x_1+x_2}$	$\dfrac{e_1+e_2}{x_1+x_2}\times 100$	Correct

$$
\begin{array}{llll}
x_1 - x_2 & e_1 + e_2 & \dfrac{e_1 + e_2}{x_1 - x_2} & \dfrac{e_1 + e_2}{x_1 - x_2} \times 100 \quad \text{Correct} \\[3mm]
x_1 x_2 & x_1 e_2 + x_2 e_1 & r_1 + r_2 & p_1 + p_2 \qquad\qquad \text{First Approx.} \\[3mm]
\dfrac{x_1}{x_2} & \dfrac{x_1}{x_2}\left(\dfrac{e_1}{x_1} + \dfrac{e_2}{x_2}\right) & r_1 + r_2 & p_1 + p_2 \qquad\qquad \text{First Approx.} \\[3mm]
& = \dfrac{x_1}{x_2}(r_1 + r_2)
\end{array}
$$

The method of approximations can be extended to operations containing more than two quantities. Thus,

$$
\begin{aligned}
x_1 x_2 x_3 &\simeq (x_1 + e_1)(x_2 + e_2)(x_3 + e_3) \\
&\simeq x_1 x_2 x_3 (1 + e_1/x_1)(1 + e_2/x_2)(1 + e_3/x_3) \\
&\simeq x_1 x_2 x_3 (1 + e_1/x_1 + e_2/x_2 + e_3/x_3)
\end{aligned}
$$

Therefore, the approximate error

$$
\simeq x_1 x_2 x_3 (e_1/x_1 + e_2/x_2 + e_3/x_3)
$$

the relative error $\simeq e_1/x_1 + e_2/x_2 + e_3/x_3 = r_1 + r_2 + r_3$
and the percentage error $\simeq p_1 + p_2 + p_3$

$$
\begin{aligned}
\frac{x_1 x_2}{x_3} &\simeq \frac{(x_1 + e_1)(x_2 + e_2)}{x_3 - e_3} \\[2mm]
&\quad \frac{x_1 x_2 (1 + e_1/x_1)(1 + e_2/x_2)}{x_3 (1 - e_3/x_3)} \\[2mm]
&\quad \frac{x_1 x_2}{x_3}(1 + e_1/x_1)(1 + e_2/x_2)(1 + e_3/x_3) \\[2mm]
&\quad \frac{x_1 x_2}{x_3}(1 + e_1/x_1 + e_2/x_2 + e_3/x_3)
\end{aligned}
$$

The approximate absolute error $\quad \simeq \dfrac{x_1 x_2}{x_3}\left(\dfrac{e_1}{x_1} + \dfrac{e_2}{x_2} + \dfrac{e_3}{x_3}\right)$

The approximate relative error $\quad \simeq \dfrac{e_1}{x_1} + \dfrac{e_2}{x_2} + \dfrac{e_3}{x_3}$

$\simeq r_1 + r_2 + r_3$

The approximate percentage error $\simeq p_1 + p_2 + p_3$

Example
Calculate, to a first approximation, the relative error in

$$
\frac{12 \cdot 4 \times 0 \cdot 15}{6 \cdot 3 \times 1 \cdot 29}
$$

if the numbers are given in significant figures.

Quantity	Absolute error	Relative error
12·4	0·05	0·004
0·15	0·005	0·033
6·3	0·05	0·008
1·29	0·005	0·004
	Total relative error	0·049
		$\simeq 0\cdot05$

Example

If 3, 2 and 4 are the largest probable errors in 230, 193 and 526, respectively, calculate the approximate largest percentage error in

(i) $230 + 193 + 526$

(ii) $526 + 230 - 193$

(iii) $230 \times 193 \times 526$

(iv) $\dfrac{230 \times 193}{526}$

(v) $\dfrac{230}{526 \times 193}$

(vi) $\dfrac{230 + 193}{526}$

Number	Absolute error	Percentage error
230	3	1·304
193	2	1·036
526	4	0·760

(i)   Largest absolute error = $3 + 2 + 4 = 9$. Sum of the numbers = 949

Percentage error $= \dfrac{9}{949} \times 100 \simeq 0\cdot95$

(ii) Largest absolute error = $3 + 2 + 4 = 9$. Sum of numbers = 563

Percentage error $= \dfrac{9}{563} \times 100 \simeq 1\cdot60$

(iii) Approximate percentage error $= 1\cdot304 + 1\cdot036 + 0\cdot760 = 3\cdot10$

(iv) Approximate percentage error $= 1\cdot304 + 1\cdot036 + 0\cdot760 = 3\cdot10$

(v)   Approximate percentage error $= 1\cdot304 + 1\cdot036 + 0\cdot760 = 3\cdot10$

(vi) Approximate percentage error in $230 + 193 = \dfrac{5}{423} \times 100 = 1\cdot18$

Approximate percentage error in $\dfrac{230 + 193}{526} = 1\cdot18 + 0\cdot76 = 1\cdot94.$

Exercise 2.6

1. If 12·1, 3·4, 0·43 and 29 are given in significant figures, calculate the approximate relative and percentage errors in

(i) $12\cdot1 + 3\cdot4$ (ii) $0\cdot43 + 29$ (iii) $\dfrac{0\cdot43}{29}$

(iv) $3 \cdot 4 - 0 \cdot 43$ (v) $\dfrac{12 \cdot 1 \times 3 \cdot 4}{0 \cdot 43}$ (vi) $\dfrac{12 \cdot 1 + 3 \cdot 4}{0 \cdot 43 + 29}$

(vii) $12 \cdot 1 + 3 \cdot 4 - 0 \cdot 43 + 29$ (viii) $\dfrac{12 \cdot 1 \times 3 \cdot 4}{0 \cdot 43 \times 29}$

(ix) $\dfrac{0 \cdot 43 + 29}{12 \cdot 1}$

2. If e_n, r_n, p_n are the absolute, relative and percentage errors in x_n, calculate the approximate absolute, relative and percentage errors

in   (i) $\dfrac{x_1^2}{x_2}$ (ii) $\dfrac{4x_1}{x_2^2}$ (iii) $\dfrac{2x_1}{x_2}$

3. Calculate, to a first approximation, the absolute error in $\dfrac{x_1 + x_2}{x_3}$

4. 12 500 people showed average earnings of £10·8.
   (a) Calculate the total earnings if the number of people is given to the nearest 100 and the earnings to the nearest £0·1.
   (b) Calculate the relative error in the earnings.
   (c) Explain the difference between relative errors and unbiased errors and state the circumstances under which relative errors would be of the greatest use.   (*R.S.A.I.*)

Further Approximations
For students with a knowledge of indices, the following approximations will be useful. If a is a small quantity in comparison with 1, then

$(1 \pm a)^2 = 1 \pm 2a$; $(1 \pm a)^3 = 1 \pm 3a$; and so on.

$\sqrt{1 \pm a} = (1 \pm a)^{\frac{1}{2}} = 1 \pm a/2$; $\sqrt[3]{1 \pm a} = (1 \pm a)^{\frac{1}{3}} = 1 \pm a/3$ and so on.

$\dfrac{1}{(1 \pm a)^2} = (1 \pm a)^{-2} = 1 \mp 2a$; $\dfrac{1}{(1 \pm a)^3} = (1 \pm a)^{-3} = 1 \mp 3a$ and so on.

$\dfrac{1}{\sqrt{1 \pm a}} = (1 \pm a)^{-\frac{1}{2}} = 1 \mp a/2$; $\dfrac{1}{\sqrt[3]{1 \pm a}} = (1 \pm a)^{-\frac{1}{3}} = 1 \mp a/3$ and so on.

Example
If a, b and c are small quantities calculate, to a first degree of approximation, the value of

$$\frac{\sqrt{1 + a}(1 + b)^3}{\sqrt[3]{1 + c}}$$

$\sqrt{1 + a} \simeq 1 + a/2$; $(1 + b)^3 \simeq 1 + 3b$; $\sqrt[3]{1 + c} \simeq 1 + c/3$.

Therefore, $\dfrac{\sqrt{1 + a}(1 + b)^3}{\sqrt[3]{1 + c}} \simeq \dfrac{(1 + a/2)(1 + 3b)}{(1 + c/3)}$

$\simeq (1 + a/2)(1 + 3b)(1 - c/3)$

$\simeq 1 + a/2 + 3b - c/3$

Exercise 2.7

If a, b and c are small quantities calculate, to the first degree of approximation, the values of

(i) $\dfrac{(1+a)^2}{(1+b)^3}$;   (ii) $\sqrt{\dfrac{1+a}{1+b}}$;   (iii) $\dfrac{\sqrt{1+a}}{(1-b)^3}$;   (iv) $\dfrac{\sqrt{1+a}+\sqrt{1+b}}{\sqrt{1+c}}$;

(v) $\dfrac{(1+a)^3 \sqrt[3]{1+b}}{\sqrt[4]{1+c}}$

3   Classification and Tabulation

Classification

Suppose that out of thirty pupils to enter the school we wish to compare the number with blue eyes with the number with brown eyes. We can make entries in a notebook and obtain, say, the following information.

Blue, brown, brown, blue, brown, blue, brown, blue, brown, blue, brown, brown, brown, brown, brown, blue, brown, brown, blue, blue, brown, brown, blue, brown, brown, blue, brown, brown, brown, blue.

It is difficult to pick out any feature from information presented in this way, but if we tabulate the information the important features become evident.

Eye Colour of Thirty Pupils

Colour	Number
Brown	19
Blue	11
Total	30

The bringing together of items with a common characteristic is known as classification. In our example we have classified the pupils according to the colour of their eyes.

There are four methods of classification. Thus the pupils in a school can be classified on a:

1. *time* basis; for example, the number of times they attend school in a term, year, and etc.
2. *geographical* basis; for example, according to the district in which they live.
3. *qualitative* basis, that is according to some attribute; for example sex, colour of hair, School House.
4. *quantitative* basis, that is according to some characteristic that can be measured; for example, height, weight.

Variable

The characteristic that varies from one member to another is known as a variable. A variable can be either continuous or discrete. For example, the height of a boy is a continuous variable, for as he grows from say

1 mm to 1·5 m, his height passes through all values between these limits; but the number of rooms in a house is a discrete variable, because it can only take certain values, such as 1, 2, 3, and so on.

Sometimes a continuous variable is treated as if it were discrete. If we measure the heights of pupils in a school correct to the nearest centimetre, the variable is treated as if it were discrete with a difference of a centimetre between the values of the variable, and a discrete variable whose discontinuities are small in comparison with the size of the variable is treated as if it were continuous; for example the annual population of a country.

Tabulation

After material has been collected and classified it is necessary to arrange it in a small space so that the eye can easily take it in, see at a glance the important results, and if possible, discover a connection between the factors. This is the purpose of tabulation.

Many examples of well tabulated material can be found in publications of Her Majesty's Stationery Office (*Monthly Digest of Statistics*, *Education*, *Annual Abstract of Statistics*, etc.) and these should be carefully studied. If a completed table looks simple and gives the appearance of being easy to construct it is because it has succeeded in its purpose. But do not fall into the error of assuming that tables are always easy to construct. Much work is needed to collect and classify the subject-matter of the table, and a large amount of practice is needed to become proficient in the construction of tables.

Examples are given below of tabulation varying from simple to complex.

(a) *Simple Tabulation* This consists of entering in columns characteristics stated at the head of the column.

Y-Shire School Pupils Classified in Forms	Form	Number of Pupils
	1	33
	2	32
	3	30
	4	31
	5	29
	6	12
SOURCE: School Register	Total	167

Y-Shire School Pupils Classified in Forms, Sex and Age

Age on September 1st 1978

Form	11 yrs.			12 yrs.			13 yrs.			14 yrs.			15 yrs.			16 yrs.			17 yrs. and over			Total		
	Boys	Girls	Total	Boys	Girls	Total	Boys	Girls	Total	Boys	Girls	Total	Boys	Girls	Total	Boys	Girls	Total	Boys	Girls	Total	Boys	Girls	Total
1	3	2	5	18	10	28																21	12	33
2				4	2	6	16	10	26													20	12	32
3							2	1	3	17	10	27										19	11	30
4										1	1	2	19	10	29							20	11	31
5													2	1	3	17	9	26				19	10	29
6																2	1	3	6	3	9	8	4	12
Total	3	2	5	22	12	34	18	11	29	18	11	29	21	11	32	19	10	29	6	3	9	107	60	167

SOURCE: School Register

(b) *Complex Tabulation* This involves more than one characteristic. The following table shows the age of pupils and the size of form, and uses both columns and rows.

Y-Shire School Pupils Classified in Forms and Age Groups

Form	Age on September 1st 1978							Total
	11	12	13	14	15	16	17 and over	
1	5	28						33
2		6	26					32
3			3	27				30
4				2	29			31
5					3	26		29
6						3	9	12
Total	5	34	29	29	32	29	9	167

SOURCE: School Register

(c) *Further Complex Tabulation* This can be achieved by sub-dividing the columns or rows to illustrate further characteristics.

The table opposite tabulates three characteristics, age, sex of pupils, and size of form.

Skill and practice are required to tabulate material so that the relationships between the quantities become obvious. Rules to cover every particular case cannot be given, but the following general rules should always be obeyed. The pupils is advised to learn them.

Rules for Tabulation
1. Be neat in all work.
2. Classify the material; i.e. decide the compartments into which the material is to be divided.
3. Make a rough draft, giving consideration to the layout of the rows and columns, and to the final appearance of the table.
4. Title all tables.
5. Name all columns and rows.
6. State all units of measurement. If the numbers are large, give the units in hundreds, thousands, etc.
7. Underline all totals and other important results, or use a variation in type, so that they are obvious to anyone looking for them.

29

8. Use light rulings to separate sub-columns, and heavier rulings, or double lines, to distinguish main columns.
9. Give the source of all material used.
10. Place a note at the bottom of the table if any heading or figure needs further clarification; a symbol may be used to draw attention to it.
11. Explain the meaning of all symbols used.

Illustration

The foregoing general rules are illustrated by the following table, published by the Department of Energy, giving the production of electricity in Great Britain. Rules 1 and 2 have obviously been obeyed, while Rule 3 is illustrated in the vertical layout of the rows. Extra spacing has been given at suitable spaces to lengthen the table, and make the width and length more in keeping with each other. Rules 4, 5, and 6 are illustrated, and as the numbers are large, the electricity units are in units of one thousand million watt-hours and the fuel in thousand tonnes. This also eliminates unnecessary width in the columns by omitting the noughts. Thicker type has been used to give the total electricity generated, whilst light and heavy rulings have been used to distinguish main columns and the source of the information is given.

Production of Electricity in Great Britain 1970–75

	1970	1971	1972	1973	1974	1975
	GWh					
Electricity Generated						
Total	**228900**	**236417**	**243438**	**259503**	**251115**	**251952**
Public Supply	228236	235750	242745	258800	250466	251263
Railways and Transport	664	677	693	703	649	689
Methods of Generation						
Steam plant (Nuclear)	21870	23209	25304	23658	29395	26518
Steam plant (Other)	200532	208485	212786	230595	216350	220301
Gas Turbines and oil engines	1517	969	1611	1347	1139	779
Hydro-electric plant	3856	2844	2865	3235	3534	3201
Pumped storage plant	1125	910	872	668	697	1153
	Thousand Tonnes					
Fuel used						
Coal	76047	71612	65009	75306	65800	73150
Coke and Coke breeze	122	67	43	64	69	134
Oil	12387	14524	18575	16708	16922	12616
Natural gas	235	1031	2472	1118	3862	3365

SOURCE: Dept. of Energy

Much practice is needed to draw up a successful statistical table and great care is needed to avoid mathematical errors. The reader is advised to study published statistical tables, and criticise their presentation.

Exercise 3.1

1. Describe briefly the first steps in tabulation, describing the various methods.

2. Define what is meant by (*a*) a continuous variable, (*b*) a discrete variable.

3. Describe briefly the purposes of tabulation.

4. State general rules for tabulation.

5. Classify the following into continuous and discrete variables:
   (*a*) Number of persons killed on the roads per week.
   (*b*) Marks awarded in a test.
   (*c*) Daily barometric heights in London.
   (*d*) Number of words in a sentence.
   (*e*) Number of persons killed annually on the railways.
   (*f*) Air temperature.
   (*g*) Number of frosty days per year in London.
   (*h*) A boy's weight.

6. Construct blank tables for the following analysis of the pupils in a school, paying careful attention to the rules of tabulation, and showing all totals. (Assume the school is co-educational.) The pupils classified in:
   (*a*) Form and House.
   (*b*) Form and Sex.
   (*c*) Form, House, and Sex.
   (*d*) Form, Sex, and Number who stay at school for lunch.

7. Draw up a table to show the results of last year's school games, (win, draw, lose), if the boys play football and cricket, and the girls hockey and tennis, and there are three schools teams for each game.

8. Design a compact table to record the number of sunspots for each year from 1850 to 1960.

9. For the annual athletic sports, a school's pupils are classified into lower, middle, and upper groups. The boys' events are 100 m, 200 m, 400 m, 800 m, high jump, long jump, pole vault, discus,

weight, javelin, hop, step and jump. The girls' events are 100 m, 200 m, high jump, long jump. There are also house Relay races for the four houses (Red, White, Blue, and Green) for both boys and girls in each group. Draw up a table to show the names of those placed first, second, and third in each event; the result compared with the school record and the points awarded to each House.

10. Design a table which could be used to show the time devoted to each class of programme broadcast during one particular week on Radios 1, 2, 3 and 4 of the B.B.C., using the following classifications: Religious Services, Music, Variety, Plays, Talks, Sport. (*L*)

11. Most state schools are classified as either primary or secondary. Primary schools are sub-divided into infant departments (5 to 6 years) and junior departments (7 to 10 years). Pupils from 11 years of age attend secondary schools which may be modern, technical or grammar schools and all pupils must remain at school until they are at least 16 years of age. Design a table to show the age distribution of the school population. (*L*)

12. New building falls into three classes—Housing, Industrial, Other. Draw up a table designed to show the value in each class of new building completed in the four quarters of the year 1976 sub-dividing each class into that undertaken for public authorities and that for private owners. (*A.E.B.*)

13. (*a*) Classify the following into continuous and discrete variables:
     (i)   The barometric heights at Kew over a period.
     (ii) Number of persons visiting the local cinema per day for a month.
     (iii) The tension in an elastic string when stretched beyond its natural length.
     (iv) The number of goals scored per match by the school football team in a season.
     (v)   The number of boys per family in a given district.
    (*b*) Design with careful attention to form and detail, and showing all sub-totals, a blank table in which could be shown a school's results in the General Certificate of Education for last year analysed into Ordinary and Advanced Level students, and the ages of students in years. The results can be 0 to 5 or more passes at Ordinary Level, and 0 to 3 or more passes at Advanced level. The youngest candidates are 14 years old and a few of the oldest candidates are over 19 years old. (*A.E.B.*)

32

14. The records of a firm show that at January 1st, 1976 the total staff employed was 100 of whom 74 were women. During that year 18 of the staff left of whom only 4 were men. The recruitment of new staff during 1976 was 3 men and 11 women.

    During the following year there were 5 less employees leaving than in 1976 and of this wastage only one man was included. The recruitment for 1977 showed an increase of 3 women and 2 men over the corresponding figures for 1976.

    Arrange this information in concise tabular form showing all relevant totals and sub-totals. (*A.E.B.*)

15. During 1960 six railway employees were killed and eighty-six injured in train accidents. Compared with 1959 this was a decrease of two in the number killed but an increase of thirty-one in the number injured.

    Of the six killed in 1960 three were engine drivers and the rest were gatekeepers. In 1959 three more engine drivers but two fewer gatekeepers were killed. The other man killed in 1959 belonged to the class known as other grades of workmen.

    In 1959 twenty-three engine drivers and nineteen guards were injured, and in 1960 these numbers were increased by twenty and five, respectively. One gatekeeper and eight permanent-way men were injured in 1960, and the other injuries in both years were amongst the other grades of workmen.

    Tabulate this information in a concise form. (*A.E.B.*)

16. Design with careful attention to form and detail, and showing relevant sub-totals, a blank table to show the number of males and the number of females serving in H.M. Forces in June for the last three years, H.M. Forces to be classified as Navy, Army, and Air Force. (*A.E.B.*)

17. Every year the Government's agricultural statistics department receives returns from agriculturists in England and Wales, Scotland, and Northern Ireland on the area of land used for agricultural purposes and the harvest obtained from it. The land is classified as (i) arable, (ii) permanent grassland and (iii) rough grazings.

    The arable land is subdivided into (i) crops and fallow, (ii) lucerne (for England and Wales only), and the permanent grassland into (i) land for mowing, and (ii) land for grazing.

    The area is given in units of five hundred hectares and the harvest in units of thousand tonnes.

    Design, showing all relevant sub-totals, a blank table to show the above information for the last three years. (*A.E.B.*)

18. Summarise the desirable properties of a table.

Draw up the framework of a table to show for 1965, 1970 and 1975 the population (to the nearest thousand) of England, of Scotland and of Wales, and to show also for each year the population of each country expressed as a percentage of the total.

(L)

19. State briefly the purposes of tabulation and give two characteristics of a good table.

Design the framework of a table to show how the number of families in each of five different income groups varies with the number of members in the family. (L)

4 Pictorial Representation

Introduction
The presentation of numerical data in the form of tables has been discussed, but most persons find the presentation far more effective and easier to understand if represented in a pictorial or diagrammatic form. In modern times the diagrammatic presentation of facts plays an important part in our lives, as diagrams can be seen every day in newspapers and advertisements. Many of them are good, and present a very fair picture, but many are misleading and can give a wrong impression to someone not skilled in their interpretation, or without a knowledge of the figures they represent.

Purposes
The purposes of pictorial representation are to make it easier to:
1. hold in mind the mass of figures they represent;
2. show more clearly than the tables can any relations which may exist between the figures;
3. show degrees of difference between items;
4. make a quick, lasting, and accurate impression of the significant facts.

It is useless using pictorial representation unless the diagrams are:
1. easier to understand than the figures they represent;
2. clear in their meaning;
3. adequately labelled, and carrying all the information necessary to appreciate them fully.

Methods of Presentation
There are many ways of representing statistical data in pictorial form, and they offer many opportunities for original ideas. The following are some of the standard methods of presentation.

Bar Charts (or Diagrams)
If the information is of the quantitative form it is usually represented by bar charts or pie diagrams. Thus the following table gives the number of km² of land devoted to the cultivation of fodder crops.

Area Devoted to Fodder Crops in Great Britain

Crop	km²
Beans	488
Peas	140
Turnips and Swedes	2296
Mangolds	824
Other Crops	2292
Total	6040

SOURCE: Agricultural Dept.

To represent this information by means of a bar chart, bars of equal width are constructed with their lengths proportionate to the area. Thus if 1000 km² were represented by a bar 2·0 cm long, beans would be represented by a bar 0·98 cm long, peas by a bar 0·28 cm long, and so on. A scale is placed at the side of the chart, so that some idea of the

Area devoted to fodder crops in Great Britain

magnitude of each item is obtained. The diagram has a neater appearance if boxed, and if the information is of such a nature that it can be rearranged to allow the bars to be in ascending or descending order of magnitude.

The judicious use of colour or shading adds to the attractiveness of the chart. The shading should be light and even, and soft pastel colours may be considered more pleasing than brilliant hard ones.

To show differences
If the bars are of approximately the same length, and it is necessary to show the slight differences in the magnitude of the quantities, it is usual to break the bar and use a large scale. It is essential to draw attention to this on the diagram.

Example
The number of pupils in a school classified in forms is:

Form	1	2	3	4	5
Number of pupils	36	35	33	30	31

Represent by a bar chart.

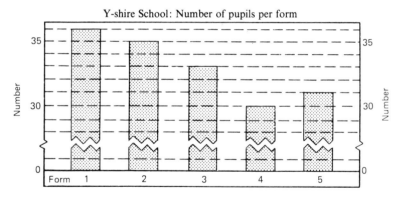

Y-shire School: Number of pupils per form

Pie Charts (or Diagrams)
In our previous example the proportion of each class to the whole can be shown by placing the bars end to end.

Turnips and Swedes	Other Crops	Mangolds	Beans	Peas

This presentation is not effective if the bar is long, for all the components cannot be seen at a glance.

An alternative method of presentation is by means of a pie chart. A circular 'pie' is divided into sectors, and the area of each sector is proportionate to the magnitude of the class it represents. The area of a sector of a circle is proportionate to the angle of the sector, and the angle can be calculated as follows. As the total area of the pie is proportionate to the total area under cultivation, we have:

$$\frac{488}{6040} = \frac{\text{Area growing beans}}{\text{Total area under cultivation}} = \frac{\text{Area of sector}}{\text{Area of circle}} = \frac{\theta^\circ}{360^\circ}$$

where θ is the angle of the sector.

Therefore $\theta^\circ = \dfrac{488 \times 360^\circ}{6040} = 29^\circ$

and the corresponding angles for the sectors representing peas, turnips and swedes, mangolds, and other crops are 8°, 137°, 49°, 137°, respectively.

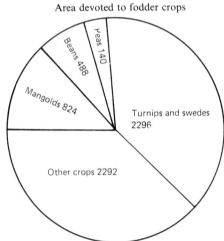

Area devoted to fodder crops

Peas 140

Beans 488

Mangolds 824

Turnips and swedes 2296

Other crops 2292

Again the figure is improved if the sectors are arranged in order of magnitude, and if colour or shading is used judiciously.

It is interesting to note that the length of the arc of a sector is also proportionate to the angle of the sector; so that in a pie diagram the proportions of the items are represented in three ways—the area, angle and arc of the sector.

Symbol Charts (or diagrams)

Another method of pictorial representation is the symbol, ideograph, pictogram or isotype chart. In this method a symbol, which can readily be associated with the class under consideration, is used to represent a unit of the class. Thus if we wish to compare the area under cultivation for beans and peas we could draw ten beans to represent the 488 km² and three peas to represent the 140 km², where each symbol represents 50 km².

Area under cultivation for beans and peas

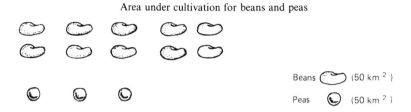

Beans ⬭ (50 km²)
Peas ⬭ (50 km²)

In this method it is not recommended to try to express the degree of accuracy to more than half a unit, as it is difficult to draw accurately fractions of a symbol.

Comparison by Lengths, Areas and Volumes
Most people find it far easier to compare lengths, so comparisons by area and volume must be judged with greater care. If any ambiguity is possible when information is given in a pictorial form, it should always be clearly stated whether the pictures are to be compared by length, area, or volume.

For example, let us compare two quantities A and B in the ratio of one to two.

(a) *By length:*

A ▭

B ▭

If the length of A is one unit then the length of B is two units.

(b) *By area:*

The area of A is to the area of B as one is to two. Therefore if the length of a side of A is one unit, then the length of a side of B is $\sqrt{2}$ units, i.e. 1·4 units.

39

(c) *By volume:*

 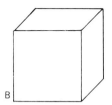

If the length of a side of A is one unit, then the length of a side of B is $\sqrt[3]{2}$ units, i.e. 1·26 units.

Most persons looking at (b) would form a comparison to which both the area of the figures and the length of the sides have contributed, and this would result in a smaller apparent difference between the items. In case (c), the differences will be further underestimated.

But, if in a diagram of an area or volume, the lengths of the sides are drawn in the ratio of one to two, the difference between the items is then exaggerated.

(a) *By area:*

 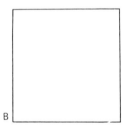

The real ratio represented in this diagram is one to four.

(b) *By volume:*

 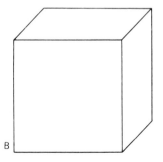

The real ratio represented in this case is one to eight.

So when numerical facts are represented in diagrammatic form involving areas or volumes it is essential, to avoid ambiguity, that it should be stated whether the figures are to be compared by lengths, areas, or volumes.

Advantages and Disadvantages of each Presentation
Bar, pie, and symbol charts all have their uses, and the advantages and disadvantages of each are listed below.

(*a*) *Bar Chart:*
1. Easy to draw.
2. Fairly accurate.
3. Comparative size of items can be shown better than with the pie chart.
4. The values of the quantities represented can be read off from the scale of values.
5. The significance of the information they represent is more easily grasped than by the figures they represent.

(*b*) *Pie Chart:*
1. Not so easy to draw as a bar chart.
2. Not susceptible to such accuracy.
3. The values of the items cannot be read off from the diagram but must be given.
4. The significance of the information they represent is more easily grasped than by the figures they represent.
5. Shows the relationship of the parts to the whole better than the bar chart.

(*c*) *Symbol Chart:*
1. Although similar to the bar chart it is not so accurate.
2. It is more appealing to persons requiring only simple and less accurate information.
3. The drawing of exactly similar repeated symbols is difficult. It is recommended that a stencil be made.

Example
Calculate the angles of the sectors in a pie diagram if quantities are represented in the proportions 476, 583, 673, 973 units.

The total area of the pie represents 2705 units.

Therefore if θ_1 is the angle for the first sector

$$\frac{\theta_1}{360} = \frac{476}{2705}$$

Number of passengers carried on the domestic air services of Great Britain

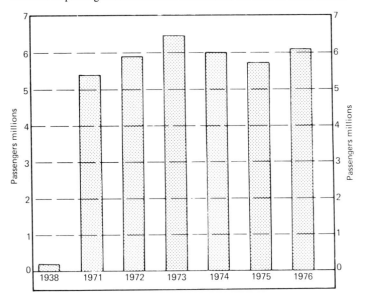

Imports of food

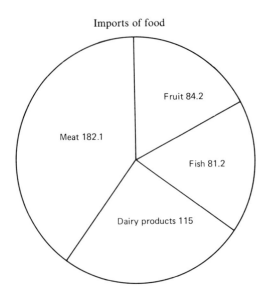

$$\text{i.e. } \theta_1 = \frac{476 \times 360}{2705} = 63°,$$

$$\text{and } \theta_2 = \frac{583 \times 360}{2705} = 78°,$$

$$\theta_3 = \frac{673 \times 360}{2705} = 90°,$$

$$\theta_4 = \frac{973 \times 360}{2705} = 129°,$$

(Check $63° + 78° + 90° + 129° = 360°$.)

Example

Construct diagrams to illustrate the following:

(*a*) The number of persons carried per year on the domestic services of the United Kingdom airways.

Year	1938	1971	1972	1973	1974	1975	1976
Passengers (Millions)	2·0	5·37	5·89	6·51	6·06	5·75	6·12

(*b*) The imports of the following foods into the United Kingdom in 1976 in £m.

Meat and meat preparations	182·1
Dairy products and eggs	115·0
Fish and fish preparations	81·2
Fruit and vegetables	84·1
Total	462·4

First decide whether the information is best represented by a bar, pie, or symbol chart.

(*a*) The purpose of the table is to compare the number of passengers carried each year, and is best represented by a bar chart. As the greatest number of passengers is 6·51 millions, a suitable scale is 1 million passengers represented by 1 cm.[1]

(*b*) In this case as well as wishing to compare the quantity of each kind of food imported it is advantageous to show the proportion of each to the whole, and the information is therefore best represented by a pie chart.

Calculate the angles of the sectors.

$$\text{Meat} \qquad \frac{182·1 \times 360}{462·4} = 141·7°$$

$$\text{Dairy products} \qquad \frac{115·0 \times 360}{462·4} = 89·5°$$

[1] The year 1938 is put in for comparison.

43

Fish	$\dfrac{81\cdot2 \times 360}{462\cdot4} = 63\cdot2°$
Fruit	$\dfrac{84\cdot2 \times 360}{462\cdot4} = 65\cdot6°$

Exercise 4.1

1. State the purpose of representing information in a diagrammatic form.

2. State what information is usually represented in bar and pie charts. Describe how to construct bar and pie charts.

3. Describe possible errors in interpretation if information is represented diagrammatically by solids.

4. Quantities in the following proportions are to be represented in a pie chart. Calculate the angles of the sectors.
   (a) 1, 1, 2.   (b) 1, 2, 3, 6.   (c) 1, 2, 2, 3, 4.
   (d) 1, 8, 15, 20, 28.   (e) 143, 276, 319.

5. List the advantages and disadvantages of bar, pie and symbol charts.

6. Draw diagrams to illustrate the following, in so far as it applies to your form:
   (a) The total teaching time devoted to each subject per week.
   (b) The number of pupils in each school house.
   (c) The number of pupils who use public transport, cycle or walk to school.

7. Construct diagrams to illustrate the following information:
   (a) The average yield per acre of wheat in Great Britain, 1967–75.

Year	1967	1968	1969	1970	1971	1972	1973	1974	1975
Yield (cwt)	33·3	28·2	32·2	33·4	35·0	33·8	34·8	39·6	34·2

   (b) The harvest yield per acre in 1975 of the following cereals.

Cereal	Wheat	Barley	Oats	Mixed Corn	Rye (Grain)
Yield (cwt)	34·2	28·7	27·5	26·8	26·1

8. Illustrate by means of a suitable diagram.
   (a) The number of days with frost at Kew Observatory. (Monthly average 1914–40.)

Month	Jan.	Feb.	Mar.	Apr.	May	June
Number of days	10·1	9·0	7·5	2·2	0·1	—

Month	July	Aug.	Sept.	Oct.	Nov.	Dec.
Number of days	—	—	0·1	1·4	6·7	9·3

   (b) Number of men killed underground in 1960:
   By falls of ground 124. Haulage accidents 69. Explosions 49.
   Shaft accidents 10. Miscellaneous 35.

Exercise 4.2

1. Explain briefly the method of construction and uses of bar and pie charts in the diagrammatic representation of statistical data. State which diagram you would use to illustrate the following kinds of numerical data, giving brief reasons for your choice:
   (a) the area in Great Britain devoted to wheat, barley, oats, potatoes and sugar beet in the year 1978;
   (b) the number of motor vehicles exported from Great Britain each month during the year 1978. (L)

2. In government and other publications, statistical information is often shown in the form of ideographs; for example, pictures of sacks of different sizes or alternatively pictures of numbers of sacks of the same size, may be used to show the amounts of wheat and flour obtained from different countries.

   Comment on the advantages and disadvantages of these pictorial systems. (L)

3. What rules should be observed when diagrams are used to present statistical data?

   State which diagrams you would use to illustrate the following kinds of numerical data, giving brief reasons for your choice.
   (a) The number of births in a borough for each month of the year 1978.
   (b) The total harvest yields of the four main cereal crops, wheat, barley, oats and rye in the United Kingdom for the year 1978.
   (c) The number of pairs of socks of different sizes issued to an infantry unit during the year 1978. (L)

4. The diagram below is an attempt to show in pictorial form the relative amounts of wheat obtained from different sources and

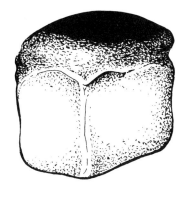

U.S.A.   Commonwealth   Home

consumed in the United Kingdom in a certain year. The actual amounts are in the ratio 7:2:1. State whether you consider that the diagram gives a clear indication of this relation, giving reasons for your opinion. (L)

5. The average weekly expenditure of households in the UK in 1975 is shown below.

Item	Expenditure (£)
Housing	7·16
Fuel, light and power	2·99
Food	13·52
Alcoholic drink	2·81
Tobacco	1·95
Clothing and footwear	4·75
Durable household goods	4·03
Other goods	4·14
Transport and vehicles	7·54
Services	5·39
Miscellaneous	0·31
	54·59

SOURCE: Dept. of Employment

Exhibit these facts in a suitable diagram.

6. State the rules that must be observed when diagrams are used to present statistical data.

    Comment on the main advantages and disadvantages of the various types of diagrams. (L)

7. The following table gives the production of steel in five districts of Great Britain in 1977:

District	A	B	C	D	E
Thousand tonnes	813	1372	2438	1880	3658

    Construct a suitable bar chart and also a pie chart of radius 5 cm to illustrate these figures. (L)

8. During an investigation into cinema-going habits, a total of 36 800 people were divided into three groups, A, B, and C. The percentages of people in the groups were, respectively, 40·5, 42·0, and 17·5. Group A was subdivided into two classes A(1) and A(2); Group B was subdivided into three classes B(1), B(2), and B(3). If the numbers of people in A(1), B(2) and B(3) were respectively, 6100, 3280, and 9720, find the number of people in Group C, and in classes A(2), B(1). Tabulate the whole of this data and illustrate by any diagram you think suitable. (L)

9. The following table shows the sources of each £1 of government revenue in 1973–74.

Source	Pence
Taxes on personal income	38
Taxes paid by companies	10
Capital gains tax and estate duty	3
Taxes on alcohol	4
Taxes on tobacco	5
V.A.T.	8
Taxes on motoring	12
Other taxation	4
Non-tax revenue	4
Borrowing	12
	100

SOURCE: National Income & Expenditure Blue Book

Illustrate the data by means of a pie chart of 5 cm radius and comment on the relative significance of each of the sources of revenue.

10. The amounts of rainfall in millimetres for the twelve months of 1975 were:

England 117, 31, 81, 71, 47, 21, 66, 52, 106, 36, 73, 53

Scotland 245, 48, 58, 100, 48, 67, 112, 86, 184, 78, 128, 88

SOURCE: Meteorological Office

Display these recordings suitably in:

(a) a table   (b) a diagram

11. (a) List the advantages and disadvantages of presenting information by means of bar charts and pie charts.

(b) The following is a list of the subjects taught in a school and the time devoted to each subject per week. Present the information in the form of a pie chart of 5 cm radius.

Subject	English	French	Mathematics	Physics	Chemistry
Time (hours)	3	3	4	$2\frac{1}{2}$	$2\frac{1}{2}$

(A.E.B.)

12. The 'invisible earnings' of the City of London in 1973 and 1974 are given below.

	£ million	
	1973	1974
Banking	112	182
Brokerage	99	159
Insurance	360	372
Merchanting	111	134
Other operations	48	58

The invisible earnings are to be represented by two comparable pie charts.

(i) Calculate to the nearest degree the angle of the sector representing each source of invisible earnings in 1973.

(ii) Using a circle of radius 5 cm, draw the pie chart for 1973.

(iii) Calculate the radius of the comparable pie chart for 1974, giving your answer correct to three significant figures. You should **not** draw the pie chart for 1974.   (*A.E.B.*)

13. The following table gives a breakdown of road accidents in 1976 by vehicle involved.

Road Accidents 1976

Pedal cycles	24 067
Mopeds	15 550
Motor scooters	2 513
Motor cycles	51 380
Cars and taxis	257 667
Goods vehicles	44 983
Other motor vehicles	24 638
	420 798

Construct a pie chart of radius 5 cm to illustrate these figures.

SOURCE: Annual Abstract of Statistics

14. (*a*) The table shows the sales of footwear in England and Wales for the years 1971 and 1974. Calculate the missing values (i) to (v) and complete the table.

(*b*) Represent the values for 1971 in a pie chart of radius 5 cm.

Type	1971	1974	% change
Men's	30·9	(i)	−25·2%
Women's	32·5	20·2	(ii)
Children's	(iii)	23·5	−20·6%
Sports shoes	3·1	3·2	+3·2%
Totals	(iv)	(v)	

Million pairs

15. The land of the United Kingdom is subdivided as follows:

England	130 359 km²	Wales	20 761 km²
Scotland	78 764 km²	N. Ireland	14 118 km²

Construct (*a*) a pie chart, (*b*) a vertical bar chart, to illustrate these figures.   (*L*)

16. In one month the sales of a baker's shop were as follows:
White Bread £260. Wholemeal Bread £40. Cakes £100.
Flour £52. Biscuits £28.

Construct a bar chart and a pie chart to represent these sales. Suggest another way by which they could be represented.

(L)

17. Write an account of circular diagrams and pictorial illustrations, explaining their advantages and disadvantages. (J.M.B.)

18.

	Tonnes	
	1975	1978
Iron	16,600	20,300
Nickel	1,750	1,690
Manganese	830	1,040
Titanium	180	396
Totals	19,360	23,426

The amounts and types of ores handled by a refinery during the years 1975 and 1978 are shown in the above table. If the total quantity used in 1975 is represented by a square having sides 4·4 cm long, calculate the size of square to represent the total for 1978.

If the constituent parts of the total for 1978 are represented in the square by rectangles of length the same as the side of the square, calculate their widths.

5 Further Examples of Pictorial Representation

Guides to Pictorial Representation

Pictorial or diagrammatic representation offers ample opportunity for the use of the imagination. Before proceeding with any problem, decide what is the important fact to be represented, and never try to represent too much information in one diagram. Always choose the simplest form of representation that allows full expression of the facts. Colour or shading should be used to emphasise some aspect of the information, and to make the diagram more attractive. The following examples illustrate common methods of diagrammatic representation.

Change Charts

Change is measured either from a 'horizontal' line where an increase is represented by a bar drawn upwards, and a decrease by a bar drawn downwards, or from a 'vertical' line where an increase is represented by a bar drawn to the right, and a decrease by a bar drawn to the left.

Example

The following table gives the number of employees in certain manufacturing industries in two consecutive years. Draw a diagram to show the changes in the number of employees in each industry over this period.

Number of Employees in Manufacturing Industries

		Manufacturing Industries							
		Employees						Thousands	
	Total	Metal Manu-factures	Engineer-ing, etc.	Vehicles	Other Metal Goods	Textiles, Leather, Clothing	Food, Drink, Tobacco	Wood, Paper, Printing	Other
1st yr.	8677	618	2032	912	545	1460	806	888	1416
2nd yr.	8794	630	2121	891	558	1461	817	899	1417

SOURCE: Ministry of Labour

Including employers and self-employed.

50

Subtracting the figures for the 1st year from those for the 2nd year we have the changes as:

Total	Metal Manu- factures	Engineer- ing, etc.	Vehicles	Other Metal Goods	Textiles, Leather, Clothing	Food, Drink, Tobacco	Wood, Paper, Printing	Other
+	+	+	−	+	+	+	+	+
117	12	89	21	13	1	11	11	1

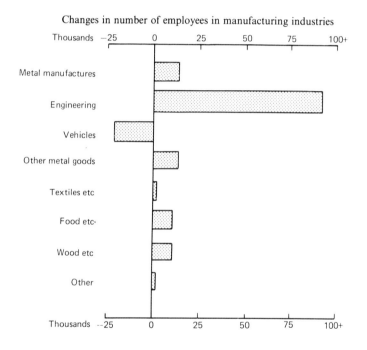

Changes in number of employees in manufacturing industries

The changes in this chart have been drawn from a vertical line as it gives a far neater appearance to the chart. It allows the proportions of the chart to be pleasing to the eye, and the names of the industries to be shown in a far more convenient way. If the changes were drawn from a horizontal line the different sizes in the names of the industries would make it difficult to space the bars neatly and equally.

Note the scale is given at the top and bottom of the chart for the greater convenience of the reader.

51

Dual Bar Charts

Generally used to show two related quantities.

Example

Number of Television Licences and Weekly Average
Admissions to the Cinema

Year	Television Licences	Weekly Average Number of Cinema Admissions
	Millions	
1956	7	21
1957	8	18
1958	9	14
1959	10	12
1960	11	10
1961	12	9

SOURCE: Annual Abstract of Statistics

Number of television licences and weekly average admissions to the
cinema

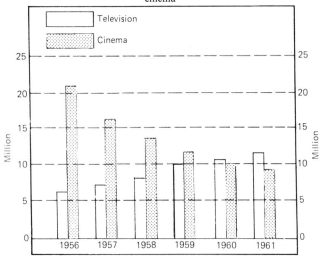

This form of presentation may be extended to Treble Bar Charts, but
it is not advisable to extend it any further.

As a point of interest, note that the comparison is between the number
of television licences issued and the weekly average number of cinema
admissions, and if we assume that on the average two and a half persons
view every television-set, then the total number of persons viewing

television and the cinema remains fairly constant over the years at about thirty-eight million.

Sectional Bar Charts
This type of chart is used to show the magnitudes of an item and its constituent parts.

Example

U.K. Exports to EEC, Oil Exporting Countries and North America (£M)

Year	EEC	OEC	N. America	Totals
1971	2659·9	578·9	1443·6	4682·4
1972	2940·3	644·7	1599·7	5184·7
1973	4034·1	800·2	1937·4	6771·7
1974	5516·2	1210·5	2262·6	8989·3
1975	6417·3	2277·6	2329·1	11024·0
1976	9168·2	3146·7	3097·1	15412·0

UK exports to EEC, oil exporting countries and North America (£M)

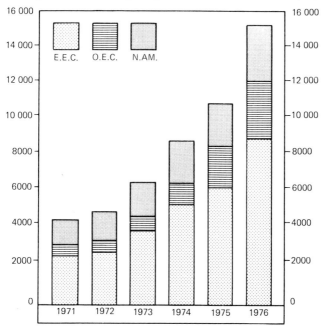

Percentage Bar Charts

The bars are divided in proportion to the percentages that the constituent parts bear to the whole. Since the scale is a percentage scale all the bars will be of the same length. Thus, the information given in the preceding example would appear as follows in a percentage chart.

Percentage chart of UK exports to EEC, OEC and North America

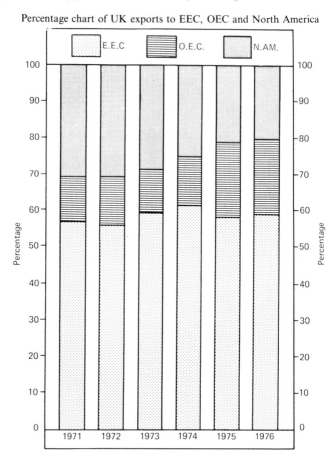

The Array

If a large number of items is to be compared and they are arranged in magnitude, and single lines, or thin bars, are drawn horizontally from a vertical line with the length of the lines proportional to the magnitude of the items they represent, the items are then arranged in an array.

An array rearranges items but does not summarise. If the values of the magnitudes of the items are arranged in order of magnitude and tabulated, the table is called a Tabular Array. Although the table or diagram gives us information which is not readily available from the original material, it is awkward to make mathematical deductions from it, as all the items are individually listed. The advantage of the array is the ease with which the magnitude of the middle, or any other item, may be read. This is important when we wish to find a special average known as the median (see page 125).

Example

Fishing stations in England and Wales that land 500 tonnes or more fish

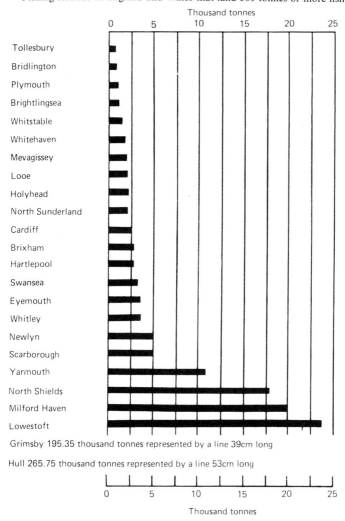

Grimsby 195.35 thousand tonnes represented by a line 39cm long

Hull 265.75 thousand tonnes represented by a line 53cm long

55

The Pyramid

This is two bar charts, one to the right and one to the left of the vertical line with corresponding groups shown opposite each other. It gives three items of information—the magnitude of each item in each of the two sets, and the total magnitude of pairs of corresponding items.

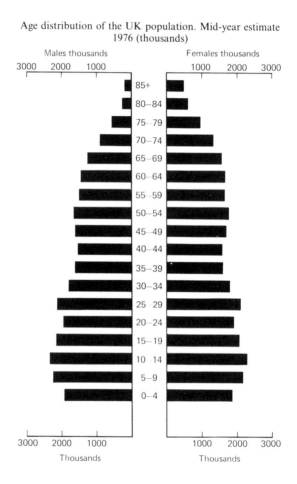

Age distribution of the UK population. Mid-year estimate 1976 (thousands)

Multi-Pie Charts

If items made up of similar constituent parts are to be compared, it is usual to draw 'pies' whose areas are proportionate to the magnitude of the items. Each pie is divided into sectors to show the proportion of the constituent items.

Example

Pelagic[1] Fish Caught at the Principal Fishing Regions
in units of 1000 tonnes

Fish	South of Ireland	English Channel	West of Scotland	North Sea
Herring	2	2	43·5	89
Mackerel	0·5	0·5	1·5	1
Pilchards		4·5		
Sprats		1		6
Total	2·5	8	45	96

[1]Fish living near the surface of the sea. Fish living in the depths of the sea are demersal fish.

The areas of the pies will be proportionate to 2·5, 8, 45, 96, or 1, 3·2, 18, 38·4, and therefore the radii of the pies will be proportionate to the square root of these numbers, i.e. to 1, 1·8, 4·2, and 6·2.

The angles of the sectors in the pies are given in the following table.

Fish	South of Ireland	English Channel	West of Scotland	North Sea
Herring	288	90	348	334
Mackerel	72	23	12	4
Pilchards		202		
Sprats		45		22
Total	360	360	360	360

The diagram is a good example of a large amount of information being presented in a manner that can be easily read and understood. Its disadvantage is the lack of precise values for the individual items, but this could be overcome by a combination of the diagram and the original table.

Quantity of pelagic fish of British taking landed from principal fishing regions

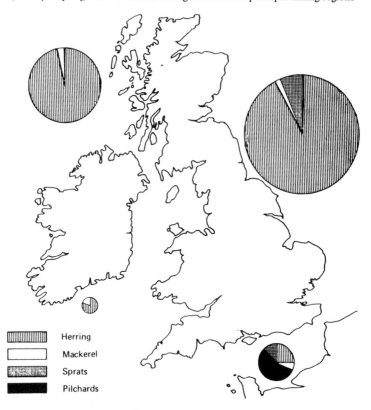

Herring
Mackerel
Sprats
Pilchards

Exercise 5.1

1. Illustrate the monthly change in the value of a commodity.

1976			1977		
	July	£235		January	£255
	August	£254		February	£257
	September	£239		March	£264
	October	£275		April	£287
	November	£278		May	£300
	December	£275		June	£264

2. Illustrate diagrammatically the change in the official reserves of the United Kingdom from the following table.

	Official Reserves (£ Million)
1968	1009
1969	1053
1970	1178
1971	2526
1972	2404
1973	2787
1974	2890
1975	2683
1976	2426

SOURCE: Annual Abstract of Statistics

3. The following table gives the current balance of the United Kingdom Balance of Payments. Illustrate diagrammatically.

	Current Balance (£ Million)
1967	−300
1968	−287
1969	+440
1970	+695
1971	+1058
1972	+105
1973	−922
1974	−3565
1975	−1701
1976	−1405

SOURCE: Annual Abstract of Statistics

4. The following in £ million are the exports of the United Kingdom to the countries named. Illustrate diagrammatically.

	1972	1973	1974	1975
E.E.C.	2940·3	4034·1	5516·2	6419·5
U.S.A.	1220·1	1524·9	1777·1	1789·1
Canada	380·0	414·0	488·1	538·5

SOURCE: Dept. of Trade

5. Represent diagrammatically the following information in what you consider to be the best way.

General Certificate of Education
Candidates' subjects at Advanced Level

	Boys	Girls	Total
17 and under	9560	3643	13203
18	16232	11035	27267
19 and over	12529	2992	15521

6. The following east coast of England fishing stations land between 5 and 500 tonnes of fish. Represent in an array.

Craster	5·6	Flamborough	16·2	Felixstowe Ferry	46·9
Amble	451·6	King's Lynn	8·3	Harwich	197·4
Newbiggin	167·7	Sheringham	9·9	Maldon	8·9
Blyth	194·5	Cromer	25·2	Southend	417·4
Seaham	5·3	Winterton	58·4	Leigh-on-Sea	84·2
Redcar	5·2	Kessingland	23·9	Margate	6·6
Staithes	29·1	Southwold	6·1	Ramsgate	43·4
Filey	173·4	Aldeburgh	62·5	Deal	7·6

(A.)

7. Represent diagrammatically the information given below.

Courses at British Universities
taken by full-time students in 1975/76

Subject Grouping	Men	Women
Education	5369	5189
Medicine, dentistry and health	17930	10001
Engineering, technology and applied science	34846	1516
Agriculture, forestry and veterinary science	3785	1361
Pure science	44433	16517
Social, administrative and business studies	38826	21397
Architecture and other professional and business studies	3984	1420
Language, literature and area studies	12316	18463
Arts other than languages, music, drama and visual arts	12098	11800

SOURCE: University Grants Committee

8. The Age Distribution in thousands for persons on June 30th, 1971 in Scotland was as follows:

Age Group	Males	Females
0– 4	228	216
5– 9	240	228
10–14	227	216
15–19	199	193
20–24	197	194
25–29	158	158
30–34	148	152
35–39	147	153
40–44	151	160
45–49	157	166
50–54	140	156
55–59	143	163
60–64	134	159
65–69	107	141
70–74	68	111
75–79	39	76
80–84	21	45
85 and over	10	27

Represent the information diagrammatically.

9. The areas of English conurbations[1] are:

	Thousand Hectares
Greater London	160
West Midlands	70
West Yorkshire	126
South East Lancashire	98
Merseyside	39
Tyneside	23
Central Clydeside	78

Show the correct proportions on a map.

10. State the rules to be observed when diagrams are used to illustrate statistical data.

Describe the types of statistical diagrams most commonly found in the national press or in advertisements and give in each case an example of the data they might illustrate. *(L)*

[1]Conurbation—Areas of urban development where a number of separate towns have grown into each other, or become linked by such factors as a common industrial or business interest or a common centre for shopping or education.

11. The numbers of thousands of live births registered in Wales, Scotland and Northern Ireland in 1969 were:

	Wales	Scotland	Northern Ireland	Total
1st quarter	11·5	25·3	8·3	...
2nd quarter	11·1	24·5	9·0	...
3rd quarter	10·8	23·2	8·4	...
4th quarter	10·1	23·2	7·7	...

Find for each quarter the total of the numbers registered. Construct a bar diagram to illustrate simultaneously these totals and the contributions to these totals from each country. (*L*)

12. (*a*) Describe briefly the difference between 'Dual' and 'Percentage' bar charts.

(*b*) The following figures give the approximate numbers in employment in the industries named for the two years 1976 and 1977. Illustrate the information in one bar chart so as to show clearly at the same time:

  (i) the total numbers employed in each industry in each of the two years,

  (ii) the difference in the numbers employed by each industry for the two years.

	Thousands	
	1976	1977
Agriculture, Forestry and Fishing	950	900
Coal Mining	650	600
Other Mining and Quarrying	70	70

13.

	Population (Millions)	
	1951	1972
England	41·2	46·3
Scotland	5·1	5·2
Wales	2·6	2·7
Northern Ireland	1·4	1·5

SOURCE: Government Statistical Service

The populations of the United Kingdom in 1951 and 1972 are to be represented by two comparable pie charts (circular diagrams)

(i) Calculate the angles at the centres of the sectors representing the populations of England, Scotland, Wales and Northern Ireland, in 1951. (Give your answer to the nearest degree)

(ii) Using a circle of radius 5 cm, draw the pie chart for 1951.

(iii) Calculate the radius of the comparable pie chart for 1972, giving your answer correct to three significant figures. You should not draw the pie chart for 1972. (*A.E.B.*)

14. (a) State briefly what kinds of data are best illustrated by the following diagrams:
    (i) Pie chart, (ii) Bar chart.
    (b) The amount of vegetables imported into the United Kingdom in the years 1953, 1957 and 1961 is given in the following table.

Thousand Tonnes

	1953	1957	1961
Potatoes	121	254	261
Onions	199	220	222
Tomatoes	185	204	158
Other	55	67	88
Total	560	745	729

Use the above table:
(i) to represent the proportions of vegetables imported in 1961 in a pie chart of radius 5 cm,
(ii) to calculate the radii of the circles which would effect a true comparison with the circle already drawn for the fresh vegetables in 1953 and 1957 if the areas of the circles are to be proportionate to the total amount of fresh vegetables imported in a year.   (A.E.B.)

15. (a) State briefly what diagrams you would use to illustrate the following data.
    (i) The number of marriages in your town for each month of 1977.
    (ii) The proportion of the total harvest yields for 1977 of each of the cereal crops, wheat, barley, oats and rye.
    (b) The following table gives the percentage composition of meadow hay harvested at different dates.

Date of Cutting	Crude Protein	Fat	Soluble Carbohydrates	Fibre	Ash
May 14	17·7	3·2	40·8	23·0	15·2
June 9	11·2	2·7	43·2	34·9	8·0
June 26	8·5	2·7	43·3	38·2	7·3

(A.E.B.)

Illustrate the above data in one diagram.

16. (a) List the purposes of diagrammatic presentation.
    (b) Represent the information given in the table diagrammatically so as to compare the amounts of food consumed in 1938 and 1963
    Comment briefly on the difference between the two years.

Consumption of Food in the United Kingdom
kg per head per annum

	1938	1963
Dairy Products	130·0	182·5
Meat and Fish	74·5	80·2
Fruit and Vegetables	165·0	178·3
Tea and Coffee	4·5	5·6
Other Foods	137·1	128·9

(*A.E.B.*)

17. Represent the information in the table in diagrammatic form to compare the number of casualties with the quarterly average new registrations.

Casualties and Road Vehicles

	Total Casualties		New Vehicles Registrations
	All Ages	Under 15 years	Quarterly Averages
Year	Thousand	Thousand	Thousand
1938	233	(i)	105
1961	350	55	315
1962	342	53	297
1963	356	56	366
1964	385	60	429

(i) Figures not given. Monthly Digest of Statistics.

(*A.E.B.*)

18. The following table is for family budgets in the United States of America and France, and shows the percentages of total income spent under each of the given headings:

	Food	Rent	Water, Heat and Light	Clothes	Health	Sundries
U.S.A.	27·7	8·1	3·5	11·0	8·2	41·5
France	60·0	4·5	3·5	9·3	4·9	17·8

Construct bar charts to illustrate these figures.

Suggest any other method by which these budgets could be compared. (*L*)

6 Frequency Distributions

Classification
In a school the heights of the boys in the same age group were measured to the nearest cm, and the following results were obtained:
167, 166, 174, 170, 168, 173, 169, 167, 172, 173, 174, 170, 167, 169, 169, 171, 173, 170, 171, 172, 170, 170, 171, 171, 172, 166, 170, 169, 165, 169, 172, 169, 174, 168, 170, 175, 173, 168, 169, 167, 169, 168, 170, 168, 174, 166, 167, 172, 176, 168, 170, 173, 171, 168, 171, 170, 169, 169, 175, 173, 167, 168, 169, 173, 170, 170, 172, 168, 170, 165, 171, 167, 172, 167, 172, 174, 173, 168, 171, 168, 174, 170, 168, 171, 174, 175, 171, 169, 166, 170, 166, 169, 172, 167, 168, 165, 173, 168, 172, 173, 169, 166, 176, 172, 166, 172, 169, 171, 169, 171, 169, 171, 169, 170

As mentioned previously, when results are presented in this haphazard manner the mind cannot grasp the important facts in the record. The figures must be classified before the facts become evident. Numerical measurements lend themselves to easy classification, for the class limits can be easily defined by values of the variable. Thus, in the example, boys of the same recorded height could be classed together, and if the results are then tabulated we get:

Heights of Boys in the same Age Group in cm, to the nearest cm.

Height	Number	Height	Number
165	3	171	13
166	7	172	12
167	9	173	10
168	14	174	7
169	18	175	3
170	16	176	2
		Total	114

Frequency Distribution
When a variable is classified according to the number of items possessing the same value of the variable, the classification is known as a frequency distribution. The interval chosen for classification (in the example one

cm) is known as the class-interval, and the number of times the items appear in a class-interval is known as the class-frequency.

In our example the class-intervals are equal, and if possible this should always be so, in order that the class-frequencies can be compared easily, but if the frequencies of some class-intervals are relatively small, it is more convenient to classify them in one larger interval.

Limits of Class-Interval

In our example the heights of the boys are measured correct to the nearest centimetre and the class-interval is one centimetre, so the heights of the seven boys classified as 166 cm will vary between 165·5 and 166·5 cm, and the heights of the nine boys classified as 167 cm will vary between 166·5 cm and 167·5 cm, and so on, that is 166, 167 cm etc. are the mid-points of the class-intervals.

There can be some ambiguity at the limits of the class-intervals as there is some doubt as to which class-interval a boy recorded as, say 166·5 cm high should belong. If the class-intervals are recorded in this manner, it is necessary to decide beforehand what to do with the items whose values coincide with the upper limit of one interval and the lower limit of the following interval. The difficulty may be overcome by measuring to further places of decimals. If the measurement proves to be 166·51 say, the height will be placed in the class-interval 166·5–167·5, if 166·49 in the class-interval 165·5–166·5. 166·5 is the *lower limit* of the one class-interval and the *upper limit* of the preceding class-interval. If this more accurate measurement is not possible then the usual method is to place half the items in each class-interval. Another alternative is to describe the class-intervals as over 165·5 cm to 166·5 cm; over 166·5 cm to 167·5 cm; and so on.

Before measurements are classified the degree of accuracy of the measurements should be given, because this affects the limiting, and the mid-point values, of the class-intervals. In our example we say the measurements are made correct to the nearest centimetre and this makes the limits of the class-interval 166 cm as 165·5 and 166·5 cm. If we say the measurements are made correct to the nearest half centimetre and the items are still classified in intervals of one centimetre, the upper limit of a sample class is 166·75 cm and the lower limit is 165·75 cm, because any measurements just greater than 166·75 are placed in the next upper class-interval, and any measurements just less than 165·75 in the next lower class-interval, and the mid-point of the class-interval is 166·25 cm. If measurements are made correct to the nearest millimetre the upper and lower limits of a sample class are 166·95 and 165·95 cm, and the mid-point of the class-interval is 166·45 cm.

Similarly in the case of ages to prevent ambiguity the date at which the ages are measured should be given. For example, if the ages of persons are measured in years on the 1st September, 1977, any person whose sixteenth birthday anniversary falls after 1st September, 1977 is classified as 15 years. It is usual to describe this type of interval as 16+, 17+, and so on, or as 16 and up to 17, 17 and up to 18, and so on. If we say the ages are given in years and months, and the class-interval is one month, any person whose sixteenth birthday anniversary falls between the 2nd and 31st of August, 1977, inclusive, is classified as 16 years 0 months, and if between 2nd and 31st July, 1977 as 16 years 1 month, and so on. If the same persons' ages are measured on any other date they will be classified differently.

In practice it is usual to select the accuracy of the measurements so as to make either the actual values of the class limits of the mid-values of the class-intervals integers or simple fractions of an integer.

This difficulty only arises in the case of continuous variables. In the case of discrete variables there is no such ambiguity. Thus the number of children in a family may be classified as 0, 1, 2, 3, etc., and no fractional value is possible. If the number of values of the variable is large, a discrete variable can still be easily classified; thus 1 to 5, 6 to 10, etc.

Histogram

To represent a frequency distribution diagrammatically, mark along the

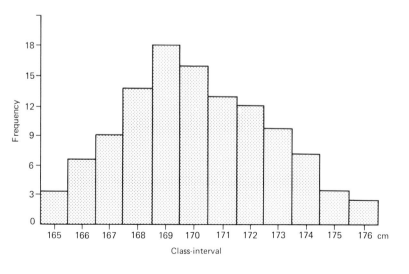

horizontal axis lengths proportionate to the class-interval and on each length construct a rectangle whose area is proportionate to the class-frequency represented by that length. In our example the class-intervals are equal, so that in this case the height of the rectangle will also be proportionate to the class-frequency.

This form of column diagram is known as a histogram.

Unequal Class-Intervals

If, in a frequency distribution the class-intervals are unequal, the areas of the rectangles above each class-interval must still be proportionate to the class-frequencies. The heights of the rectangles will therefore be proportionate to the class-frequency divided by the class-interval, as in the following example.

Class-Interval	Frequency	Height of Rectangle in Histogram
0–1 = 1 unit	10	$\frac{10}{1} = 10$ units
1–2 = 1 unit	15	$\frac{15}{1} = 15$ units
2–4 = 2 units	40	$\frac{40}{2} = 20$ units
4–10 = 6 units	120	$\frac{120}{6} = 20$ units
10–20 = 10 units	50	$\frac{50}{10} = 5$ units

Shape of Histogram

The shape of the histogram depends on the size of the class-interval. Thus, if in our example the boys are classified into two intervals from over 164·5 to 170·5, and over 170·5 to 176·5, the frequency of each class would be 67 and 47, and the corresponding histogram would be:

Size of Class-Interval

The above histogram bears little resemblance to the first, and does not give as much information. It is obvious that the size of the class-interval is important. The conditions that decide the size of the class-interval are in opposition to each other. For convenience and brevity the class-interval should be large, but it must also be small enough to allow all the values assigned to the class to be treated as if they are equal to the mid-value of the class-interval. These conditions will generally be satisfied if the number of classes is about twenty. Therefore when presented with a list of items to be classified into a frequency distribution pick out the items with the greatest and least values, divide the difference between their values by twenty and this should be an approximate value of the class-interval. The actual value should be the nearest integer or simple fraction.

Although the method described above for determining the size of the class-interval applies in the majority of problems, cases will arise where it is not applicable. For instance, if it is necessary to tabulate the number of peas in a pod, or the number of rooms in a house, we should have to use as class-intervals the integers 0, 1, 2, 3, and so on up to the largest value of the variable. In this case the discrete step is large compared with the range of variation, and no choice is possible.

Centre of Class-Interval

In the example on page 65, integral values of the variables were chosen as the mid-values of the class-intervals, but the centres of the class-intervals could have been taken as 165·5, 166·5, and so on. The items in the class-interval whose mid-value was 165·5 would have varied from over 165 to 166, and so on.

Procedure for Tabulating a Frequency Distribution

The above observations suggest that when tabulating a frequency distribution the best way to proceed is:

1. Fix the magnitude of the class-interval. If possible arrange for about 20 intervals.
2. Fix the origin of the intervals.
3. Classify the items according to 1 and 2.
4. Draw up a table showing the frequency of each class-interval.

Example

As an example of the above procedure, consider the following: in an examination the number of candidates obtaining each mark is given below. Tabulate in a convenient manner, and draw a histogram.

Table of Marks, 0 to 100

Marks	0	1	2	3	4	5	6	7	8	9
0	0	0	2	1	0	1	2	0	4	0
10	1	1	1	2	2	2	3	1	7	2
20	3	4	5	3	4	4	4	5	6	7
30	8	6	8	7	8	7	9	10	10	7
40	4	9	6	10	5	7	4	7	6	7
50	10	6	4	10	3	3	5	6	3	4
60	2	3	4	4	5	2	4	4	3	4
70	2	2	4	3	3	2	3	1	2	2
80	2	4	0	1	1	2	0	1	2	0
90	1	0	2	0	0	0	0	1	0	1
100	0									

The observations are already classified, but 101 classes are too many. The observations lend themselves conveniently to 20 classes where the class-intervals go from 1 to 5, 6 to 10, 11 to 15, and so on.

So reclassifying we obtain:

Marks	Frequency	Marks	Frequency
1– 5	4	51– 55	26
6–10	7	56– 60	20
11–15	8	61– 65	18
16–20	16	66– 70	17
21–25	20	71– 75	14
26–30	30	76– 80	10
31–35	36	81– 85	8
36–40	40	86– 90	4
41–45	37	91– 95	2
46–50	34	96–100	2

The corresponding histogram is shown below.

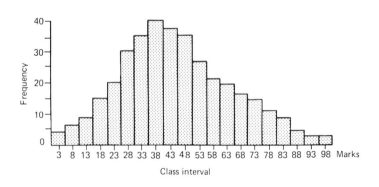

Frequency Curve

If the number of class-intervals was increased indefinitely, and each class-interval remained finite, the outline of the histogram would then approximate to a smooth curve.

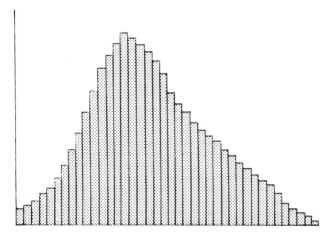

This ideal limit to the histogram is known as a frequency curve, and is of fundamental importance in statistical theory. Its fundamental property is that the area between any two ordinates is proportionate to the number of observations falling between the corresponding values of the variable.

Frequency Polygon

If we wish to compare two distributions diagrammatically it is difficult to do this with histograms plotted on the same axes. If, however, the mid-points of the tops of the rectangles in the histogram are connected by straight lines, the last point at each end being joined to the base at

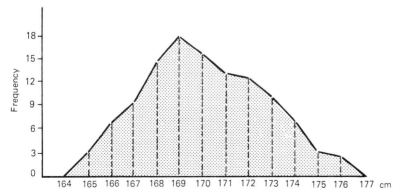

the centre of the next class-interval, the figure produced is known as a frequency polygon. Comparisons between two or more frequency distributions represented by their respective frequency polygons on the same axes are more readily distinguishable than representation by means of the corresponding histograms. The frequency polygon shown is based on the histogram on page 67.

Providing that all of the class-intervals are equal, it will be seen that the area under the frequency polygon is the same as that of the corresponding histogram because for each triangle ABC cut away from a rectangle in the histogram there is a triangle CDE equal in area added which is not in the histogram.

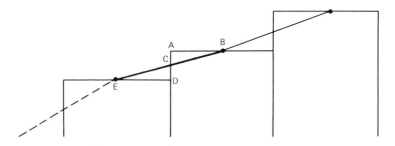

Since the area represented by the histogram is proportional to the frequency of the distribution, the area represented by the frequency polygon must likewise be proportional to the frequency of the distribution. If there are unequal class-intervals and it is important to maintain this relationship, care has to be taken to choose the correct points to join the tops of adjacent rectangles by straight lines. Thus, AB must equal CD and EF equal GH in the diagram below. Points A and H mark the centre of the top of their respective rectangles; a simple calculation will show that neither point D nor point E can be at the centre of the top of the middle triangle if AB equals CD and EF equals GH.

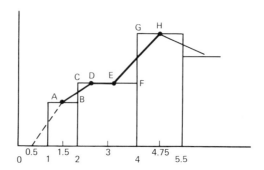

Types of Frequency Curves

The shapes of histograms and frequency curves are endless in their variety. The following are some of the common types.

1. Symmetrical Bell-shaped Distribution

The class-frequencies decrease symmetrically on either side of a maximum.

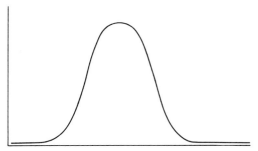

2. The Moderately Asymmetrical or Uni-modal Skew Distribution

The class-frequencies decrease more rapidly on one side of the maximum than on the other.

If the longer tail lies to the right the distribution is positively skew.

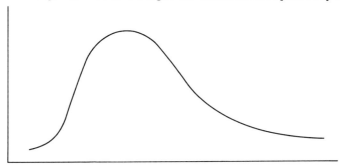

If the longer tail lies to the left the distribution is negatively skew.

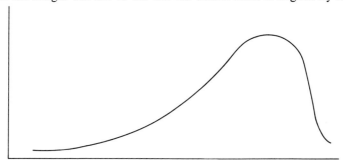

3. The Bi-modal or Two-humped Distribution

A more complex distribution that can sometimes be analysed into simpler types.

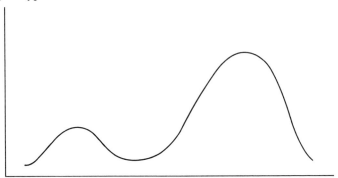

4. The Extremely Asymmetrical or J-shaped Distribution

The class-frequencies are greatest at one end of the range.

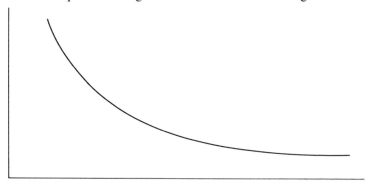

5. The U-shaped Distribution

The class-frequencies are greatest at both ends of the range, and decrease to a minimum at a point between.

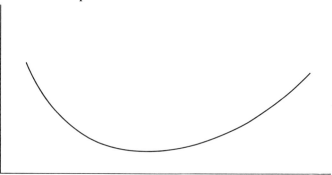

Worked Examples

1. State the mid-value of the class-intervals of the following frequency distribution, and draw the histogram.

Class-interval	0 to less than 1	1 to less than 3	3 to less than 4
Frequency	2	6	4
	4 to less than 7	7 to less than 9	9 to less than 10
	14	7	2
	10 to less than 11		
	1		

The mid-values of a class-interval can be obtained by averaging the limiting values.

Therefore the mid-values of the class-intervals are:
$$\frac{0+1}{2} = 0.5, \quad \frac{1+3}{2} = 2, \quad \frac{3+4}{2} = 3.5, \quad \frac{4+7}{2} = 5.5, \quad \frac{7+9}{2} = 8,$$
$$\frac{9+10}{2} = 9.5, \quad \frac{10+11}{2} = 10.5.$$

The class-intervals are not equal, therefore the heights of the rectangles will not be proportional to the frequencies. First calculate the height of each rectangle.

Class-interval	Frequency	Height of rectangle
0–1 = 1 unit	2	$\frac{2}{1} = 2$ units
1–3 = 2 units	6	$\frac{6}{2} = 3$ units
3–4 = 1 unit	4	$\frac{4}{1} = 4$ units
4–7 = 3 units	14	$\frac{14}{3} = 4.7$ units
7–9 = 2 units	7	$\frac{7}{2} = 3.5$ units
9–10 = 1 unit	2	$\frac{2}{1} = 2$ units
10–11 = 1 unit	1	$\frac{1}{1} = 1$ unit

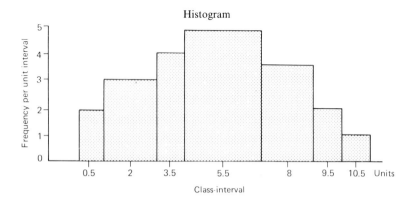

Histogram

2. State the mid-values of the class-intervals of the following frequency distribution, and draw a histogram.

Class-interval   0 1 2 3 4 5 6 7 and over.
Frequency   1 2 4 6 8 5 3 4

The mid-values of the first seven class-intervals are already given, that is, 0, 1, 2, 3, 4, 5, 6. The last interval offers some difficulty because its size is not stated. In this case, decide what to do about it. The upper limit may be 10, 12, 15, or some other relatively large number, but if the upper limit is taken to be 10, the histogram will give a fair picture. The heights of the rectangles for the first seven class-intervals will be proportionate to the frequencies, and the height of the last rectangle will be $\dfrac{4}{7 \text{ to } 10} = \dfrac{4}{4} = 1$.

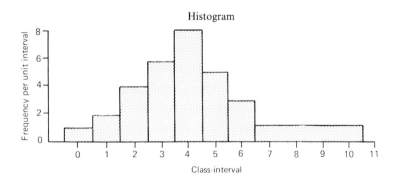

Histogram

3. Draw a histogram for the following frequency distribution.

Analysis of Agricultural Holdings by Size of Holding in England and Wales 1955–6

Size of holding (hectares)		
Frequency	0·5 and less than 2·5	2·5 and less than 7·5
	79 618	70 198
	7·5 and less than 25	25 and less than 50
	83 100	59 556
	50 and less than 75	75 and less than 150
	30 881	33 148
	150 and above	
	13 064	

The class-interval 0·5 and less than 2·5 is of two units, and its mid-value is 1·5. The class-interval 2·5 and less than 7·5 is of 5 units, and its mid-value is 5, and so on. The last interval will be satisfactory if taken from 150 to 300 ha. The heights of the rectangles for the histogram can be now calculated.

Class-interval	Frequency	Height of rectangle proportionate to:
0·5 and less than 2·5	79 618	$\frac{79\,618}{2} = 39\,809 = 39\,800$ say
2·5 and less than 7·5	70 198	$\frac{70\,198}{5} = 14\,040 = 14\,000$ say
7·5 and less than 25	83 100	$\frac{83\,100}{17\cdot5} = 4748 = 4700$ say
25 and less than 50	59 556	$\frac{59\,556}{25} = 2382 = 2400$ say
50 and less than 75	30 881	$\frac{30\,881}{25} = 1235 = 1200$ say
75 and less than 150	33 148	$\frac{33\,148}{75} = 442 = 400$ say
150 and above	13 064	$\frac{13\,064}{150} = 87 = 90$ say.

A convenient scale for the frequency is 1 cm for 4000 units. Then the heights of the rectangles are 9·95, 3·5, 1·2, 0·6, 0·3, 0·1, 0·02 cm. The last is too small to be measured, and is best represented by a thin line. If 0·1 cm represents the first class-interval, the histogram is then 15 cm long.

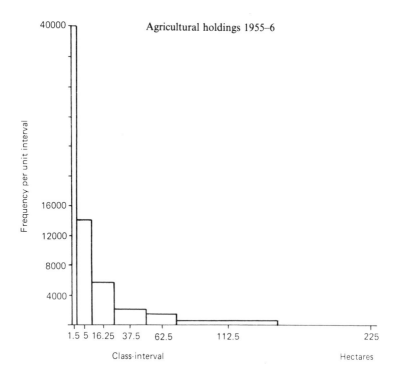

Agricultural holdings 1955–6

Frequency per unit interval

Class-interval   Hectares

Exercise 6.1

1. Define the following terms: (a) frequency distribution; (b) class-interval; (c) class-frequency; (d) histogram; (e) frequency polygon; (f) frequency curve; (g) uni-modal distribution; (h) bi-modal distribution; (i) skew distribution.

2. Describe the formation of a frequency distribution.

3. Describe the illustration of frequency distributions by histograms, frequency polygons, and frequency curves. Name the commoner types of histograms

4. Calculate the mid-values and the upper and lower limits of the class-intervals and draw histograms of the following frequency distributions:

    (a) Class-interval   1   2   3   4   5   6   7
           Frequency   1   2   4   6   3   2   1

(b)

Class-interval	0–1	1–2	2–3	3–4	4–5	5–7	7–10
Frequency	1	2	4	6	8	5	3

(c)

Class-interval	1	2	3	4	5	6	7	8	9 and over
Frequency	1	4	6	3	3	6	8	5	4

(d)

Class-interval	1–5	5–10	10–15	15–25	25–35	35–55
Frequency	10	15	24	26	24	30

(e)

Class-interval	0–10	10–20	20–25	25–30	30 and over.
Frequency	20	30	20	16	12

(f)

Class-interval	0–	5–	10–	15–	20–	25–	30–
Frequency	6	8	14	21	16	7	4

(g)

Class-interval	1–5	6–10	11–15	16–20	21–25
Frequency	9	17	22	18	10

5. With the same axes and scales draw histograms for both the male and female populations in the following frequency distributions. Name the histogram and comment on the result.

   (a) Number of marriages in the United Kingdom in 1974 classified by age and sex.

Age Group (years)	Under 21	21–24	25–29	30–34	35–44
Males (Thousand)	72	156	100	35	33
Females (Thousand)	160	130	64	25	26

	45–54	55 and over
	20	21
	17	14

   (b) Number of deaths in the United Kingdom in 1974 classified by age and sex.

Age Group (years)	0–4	5–9	10–14	15–19	20–24	25–34
Males (Thousand)	8	1	1	2	2	4
Females (Thousand)	6	1	–	1	1	2

35–44	45–54	55–64	65–74	75–84
7	26	60	111	84
5	16	34	77	112

85 and over
31
74

(– means less than $\frac{1}{2}$ a unit)

6. Draw histograms for the following frequency distributions. Name the histogram and comment on the result.
   (a) The following table gives the frequencies of estimated cloudiness at a resort for the month of July over a period of years.

Degree of cloudiness	10	9	8	7	6	5	4	3	2	1	0
Frequency	683	159	101	71	63	51	51	76	83	134	350

(b) Number of goals scored by soccer teams over a four-week period.

Goals	0	1	2	3	4	5	6	7	8	9
Frequency	243	262	177	91	52	26	9	3	4	2

(c) The height of 1165 boys of 18 years of age.

Height (cm)	165	166	167	168	169	170	171	172
Frequency	24	70	92	142	182	145	146	123

173	174	175	176	177 and over
99	74	34	18	16

7. The following marks were awarded in an examination. Re-classify and draw a histogram.

Marks	0	1	2	3	4	5	6	7	8	9
0	0	1	0	1	1	0	2	1	0	5
10	2	4	2	3	2	5	3	1	6	3
20	9	1	7	6	5	8	8	5	6	1
30	9	6	9	6	8	7	7	12	9	11
40	8	9	10	15	8	9	5	8	11	11
50	11	9	12	13	12	11	4	15	12	15
60	13	12	6	9	12	12	8	10	7	5
70	7	6	9	9	8	7	9	6	11	10
80	8	3	5	10	6	5	7	8	4	2
90	3	1	8	5	6	2	4	1	2	1
100	1									

8. The following table gives the frequencies of points scored by rugby teams over a period of two weeks. Draw a histogram. Why are there no frequencies for 1, 2, 4, and 7 points?

Points	0	1	2	3	4	5	6	7	8	9	10	11
Frequency	90	0	0	110	0	24	54	0	48	33	12	22

12	13	14	15	16	17	18	19	20	21	22	23
12	9	17	5	6	5	6	8	3	4	5	2

24	25	26	27	28	29	30	31	32	33	34	35
3	3	1	1	1	0	1	0	1	0	1	0

36
2

9. From the published results of football games find the frequency distributions of goals scored by soccer teams for two wet weeks and two dry weeks. With the same axes and scales draw histograms. Comment on the results. Repeat for rugby teams.

10. Choose a book and count the number of words in each sentence. Classify as a frequency distribution and draw a histogram. Repeat for another book by the same author, and again for a book by a different author.

    Compare and comment on the results.

    (For this exercise the students can be arranged in pairs, one doing the counting and one the recording, each pair classifying a different chapter.)

    Repeat the exercise, only this time counting the number of letters in each word. It will be sufficient to use the words in one chapter.

    (Histograms such as these are sometimes used to try to settle disputes about authorship.)

11. The following table gives the frequency distribution of the percentage marks obtained by 80 students in a mathematics examination. Construct a histogram to illustrate the data.

    It is known that the students are drawn from two classes of the same age group which are taught mathematics by the same teacher. Suggest a possible reason for any unusual feature in the histogram.

Percentage	1–10	11–20	21–30	31–40	41–50
Frequency	4	6	10	12	7
	51–60	61–70	71–80	81–90	91–100
	8	12	13	6	2

(L)

12. What is meant by 'class-interval' in frequency diagrams? State reasons why, in a frequency diagram, it is an advantage to choose a suitable class-interval, giving two examples from statistical investigations with which you have assisted. (L)

13. The following frequency distribution gives the ages of all people in the United Kingdom who were under 80 years of age on June 30th, 1975

Age	0–9	10–19	20–29	30–39	40–49	50–59	60–69	70–79
Number (Millions)	8·5	8·7	8·0	6·7	6·4	6·6	6·0	3·7

Construct the histogram and the frequency polygon of this distribution.

14. The following table gives the marks obtained by sixty candidates in Mathematics and Latin.

Marks	1	2	3	4	5	6	7	8	9	10
Maths	8	13	28	5	0	3	2	0	1	0
Latin	1	2	0	4	8	17	16	6	3	3

Draw a frequency polygon for each subject.

Comment on the advisability of getting an order of merit for the sixty candidates by adding the Mathematics and Latin marks together. (L)

15. The following table gives the marks of 110 children in an arithmetic test. Construct a frequency table for these marks, using ten class intervals 0–4, 5–9, . . . 45–50.

Hence draw the histogram and the frequency polygon for this distribution.

```
 6 10 12  1  4  5 19  4 22  6  6 23 26  2 24  6
 3  4 16  7  7 14 19 12 10 12  2  8 10 22  7 20
18 10  8 16  7 23 14  1 23 11 14 16  8 22 11  5
 2 15 22 41 42 23 15 32 25 47  5 34  9 19 31  7
30 11 25 42 19  1 17  7 27 30 43 10 41  9 13  2
10  7 31 43  6  5 17 15 10 12 16 27  6  8 12  2
16 16  9  6  2 10  7  2 23  5 10 17 19  9
```
(L)

16. The following table gives the distribution of annual incomes (after tax) between £750 and £10,000 in the U.K. for 1974–5.

Income (£)	Number of Incomes (Thousands)
750–999	2057
1000–1499	4466
1500–1999	4208
2000–2999	6343
3000–3999	2502
4000–5999	903
6000–7999	174
8000–9999	52

Construct a histogram to illustrate the table.

17. The following, in chronological order, were the annual numbers of injuries (in units of a thousand) suffered by colliery workers for the years 1924–1951 inclusive: 195, 178, 91, 173, 162, 176, 166, 141, 126, 122, 133, 134, 136, 141, 132, 134, 146, 158, 167, 174, 177, 181, 167, 163, 183, 229, 238, 234.

Taking class-intervals 90–, 120–, 150–, 180–, and 210–240, construct a frequency table for the data, and draw a histogram. Point out its most prominent features.

Discuss briefly whether the histogram alone is a fair summary of the data, and whether it would be a fair conclusion that before 1948 the mines were rarely as dangerous as they were during the period 1948–51. (*J.M.B.*)

18. An analysis of the number of words per sentence in the first hundred sentences of
(*a*) *Pride and Prejudice*, by Jane Austen,
(*b*) *The Cathedral*, by Hugh Walpole,
gives the following frequency table.

Number of words per sentence	Number of sentences	
	(a) Pride and Prejudice	(b) The Cathedral
0–4	6	2
5–9	33	12
10–14	22	14
15–19	15	19
20–24	10	18
25–29	4	6
30–34	2	5
35–39	3	6
40–44	2	6
45–49	2	4
50–54	0	1
55–59	0	1
60–64	0	2
65–69	0	0
70–74	0	1
75–79	1	0
80–84	0	0
85–89	0	0
90–94	0	1
95–99	0	0
100–104	0	2

On one sheet of graph paper draw two frequency polygons, setting them out in such a way that you can use them to compare the two distributions given above.

Discuss the use of this analysis to demonstrate that one of the novels is more difficult to read than the other. (*J.M.B.*)

19. Thirty boys took two papers in a Geography examination and their percentage marks were as follows:

Paper 1   59 59 34 54 56 43 72 40 51 48 53 52 51 59 48
Paper 2   56 74 36 54 81 39 67 62 42 62 43 83 77 55 71
Paper 1   58 48 64 42 57 65 58 34 45 65 41 39 46 57 57
Paper 2   49 42 58 42 58 52 63 35 48 54 43 29 43 54 73

Using class-intervals 10–19, 20–29, 30–39, etc., form two separate frequency distributions for the marks in Paper 1 and Paper 2, respectively and superpose in one diagram the two frequency polygons. Use your diagram to compare and contrast the two distributions. (*J.M.B.*)

20. Calculate the mid-value of each of the class-intervals, and draw a histogram for the following frequency distribution.

Class-interval	1–5	6–10	11–15	16–25	26–35	36–55
Frequency	10	15	25	60	40	40

(*A.E.B.*)

21. (*a*) Draw the outlines of the following histograms:
(i) Symmetrical, (ii) Positively skew, (iii) Negatively skew, (iv) Bi-modal, (v) J-shaped, (vi) U-shaped.
(*b*) On the same diagram draw histograms for the two age distributions given in the table. Compare and contrast the variations in the two distributions.

Population of the United Kingdom
30th June, 1975
(Thousands)

Age Group	Males	Females
0–9	4345	4108
10–19	4483	4268
20–29	4072	3963
30–39	3402	3310
40–49	4224	3222
50–59	3186	3375
60–69	3750	3259
70–79	1396	2263
80 and over	376	961

(Monthly Digest of Statistics)

22. In a savings group there are 450 members and the numbers of £1 savings units held by them are shown in the table. Construct a histogram for the distribution.

Number of units	1–50	51–100	101–150	151–200
Number of members	10	16	30	40

	201–300	301–400	401–500	501–1000
	130	90	84	50
	Total 450			(A.E.B.)

23. The table gives the monthly wages of 76 employees.

Monthly wages (£)	150+	180+	210+	240+	270+	300+
Frequency	9	12	18	16	10	8

	360–540	
	3	Total 76

State the mid-value of each class-interval of the distribution and draw the histogram.

24. Describe briefly a statistical method which you would use to investigate which of two writers appears most likely to be the author of an unidentified passage which is quoted in a newspaper, each writer having a number of books which are available for reference. *(L)*

25. Pupils on the registers in each age-group in Grant-aided and Independent Schools in January 1958.

	Age last birthday	Number of pupils in Thousands (nearest thousand)
Nursery	2– 4	211
Primary	5–10	4214
Secondary	11–15	2928
Advanced Secondary	16–18	180

Represent the above data as a histogram. *(J.M.B.)*

7   Graphical Representation and Ratio Scales

Dependent and Independent Variables
If asked to plot the graph of, say, $y = 4x^2 + 3x + 2$ the method is to give suitable values to x and calculate the corresponding values of y. As the value of y is dependent upon the value of x, y is known as the dependent variable, and x is known as the independent variable. The points are then plotted, the independent variable along the 'horizontal' axis and the dependent variable along the 'vertical' axis. A smooth curve is drawn through the points because y is a continuous function of x, i.e. for any value of x there is a corresponding value of y.

In statistics great use is made of graphs, but the quantities plotted are mostly discrete, and seldom can a smooth curve be drawn through the points. Data classified in groups, or as a time series, lend themselves to graphical representation, and, as in the algebraical example, the independent variable is plotted along the 'horizontal' axis, and the dependent variable along the 'vertical' axis. When the data forms a time series the graph showing the relationship is known as a historigram. Do not confuse this with histogram.

The method is best illustrated by examples.

Example
Represent graphically the following information:

Weekly Average Production of Steel 1977–8
(Thousand Tonnes)

Year	Jan.	Feb.	Mar.	Apr.	May	June
1977	449	425	411	380	391	390
1978	334	396	434	441	403	412

Year	July	Aug.	Sept.	Oct.	Nov.	Dec.
1977	368	371	442	403	390	292
1978	368	268	415	432	—	—

The dependent variable is the amount of steel produced. So to represent the data graphically mark along the horizontal axis equal lengths to represent the months. Along the vertical axis mark a scale to represent the amount of steel produced. As the lowest figure to be represented is 268 thousand tonnes, break the vertical scale as in drawing bar charts. *Average values of a dependent variable are plotted at the mid-point of the period they represent.* Each point represents the average weekly production of steel in a particular month, and any point between the plotted points has no meaning. Therefore no smooth curve

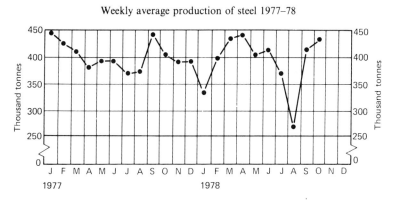

Weekly average production of steel 1977–78

is drawn through the points. As an aid to sighting the points, and to indicate the trend (i.e. whether the weekly average production of steel is increasing or decreasing), the points are joined by straight lines. But, it is stressed, only the plotted points have a meaning, and no value of the dependent variable corresponds to any point on the straight lines.

Example

Stocks of Pig-iron 1977–8 (Thousand Tonnes)
End of Period

Year	Jan.	Feb.	Mar.	Apr.	May	June
1977	844	818	741	759	747	767
1978	550	539	525	575	562	441

Year	July	Aug.	Sept.	Oct.	Nov.	Dec.
1977	788	828	794	779	707	704
1978	535	455	412	—	—	—

This example is similar to the last except that *the totals are given for the end of the period. In this case the points are plotted at the end of the period they represent.*

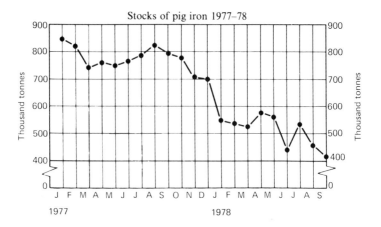

Stocks of pig iron 1977–78

Example

	Barometric Reading						
Day	1	2	3	4	5	6	7
Pressure (mm of mercury)	1001	1003	1004	1003	1005	1008	1011
	8	9	10	11	12	13	
	1020	1021	1023	1024	1025	1024	

To represent this graphically proceed as in the first example. Although barometric pressure is a continuous variable no smooth curve can be drawn through the points because the barometric pressure is erratic, and no indication is given of any intervening fluctuations. The points are again joined by straight lines, which in this case imply continuity but ignorance of the value of the fluctuations.

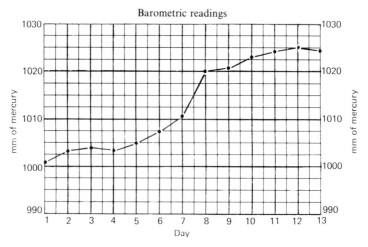

Barometric readings

Example

A vessel containing boiling water is allowed to cool, and the temperature at the end of successive minutes is taken.

Time in minutes	0	1	2	3	4	5	6	7
Temp. °C	100	88·5	82·6	77·5	72·3	68·0	64·0	60·5

	8	9	10	11	12	13	14
	57·8	55·0	52·8	50·0	48·5	46·5	45·8

The decrease in temperature is continuous, and it is unreasonable to expect fluctuations between one reading and the next. So it is permissible to draw a smooth curve through the points.

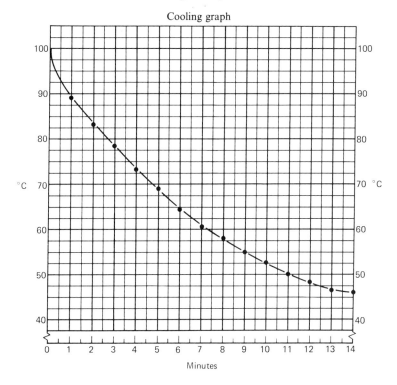

Cooling graph

Example

Population of England and Wales at each census from 1811 to 1971 in units of 1 000 000.

Year	1811	1821	1831	1841	1851	1861	1871	1881	1891
Population	10·2	12·0	13·9	15·9	17·9	20·1	22·7	26·0	29·0

	1901	1911	1921	1931	1951	1961	1971	
	32·5	36·1	37·9	40·0	43·8	46·1	48·8	(*A*.)

Population is a discrete variable, and in this case the figures quoted are correct to the nearest 100 000. We have no knowledge of intervening fluctuations, but it is again reasonable to assume that no great fluctuation occurs in any intercensal period and so the points can be joined by straight lines. But the joining by straight lines approximates to a curve, and in this case no great error is introduced by drawing a smooth curve through the points. This can always be done in the case of a discrete variable if no fluctuations are expected between values of the variable, and if the discontinuities are relatively small.

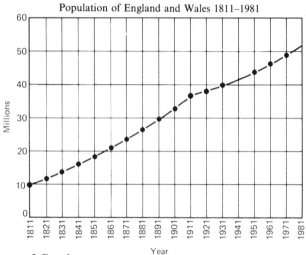

Population of England and Wales 1811–1981

Purposes of Graphs

In statistics graphs are used to:

1. present an argument pictorially; that is, to appeal to the reason through the eye. They indicate tendencies far more quickly and more convincingly than do two or three pages of figures. Thus in the last example it is noticed that on the scale of the graph the points from 1811 to 1861 lie approximately on a straight line, showing that the absolute intercensal increase in the population is constant, but from 1861 to 1911 the graph is steeper, showing that the increase in the population is greater. From 1911 to 1951 the increase seems to have again become fairly steady, and, as this part of the graph is parallel to the first part, the absolute intercensal increase in the population is the same. From 1951 the increase is rising again.

2. interpolate values of a variable when others are known. Thus the population of England and Wales is known with a high degree of accuracy at the time of each census. The population for any year

during an intercensal period can be estimated fairly accurately from the graph. Thus in 1905 the population was 34 million. To estimate when the population was, say, 25 million, read the year corresponding to a population of 25 million, which is 1878. The graph can also be used to predict populations on the assumption that no great change takes place in the intercensal increase. Thus, continuing the graph to 1981, the estimated population is approximately 51·5 million. No census was taken in 1941, but from the graph we get the estimated population of England and Wales in 1941 as 42 million.

3. suggest correlation, or connections between variables or events. This is best illustrated by examples.

Example
Mail carried on Domestic Services of United Kingdom, and Notes and coin in Circulation.

Year	1972	1973	1974	1975	1976
Mail (Thousands)	165	354	1044	1773	1681
Notes & coin (£ Million)	4079	4377	5085	5903	6714

If the values of the two variables are plotted on the same graph, they can have the same horizontal axis, because they are for the same period,

Mail carried on domestic services of United Kingdom airlines, and notes and coin in circulation

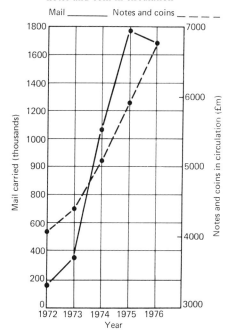

but two vertical scales must be used for the two dependent variables. The two scales are marked down either side of the graph, but the values in both cases are plotted from left to right. A suitable horizontal scale is one cm for each year, and vertically a cm for 100 thousand for mail, starting at zero, and a cm for £200 million starting at £3,000 million. In this way the relation between the two sets of figures is clearly illustrated.

Examples

Registered Unemployed and Employment Vacancies Unfilled in Great Britain 1978

	Registered Unemployed (Thousands)	Employment Vacancies Unfilled (Thousands)
January	1549	159
February	1509	172
March	1461	186
April	1452	204
May	1387	216
June	1446	228
July	1586	219
August	1608	214
September	1518	233
October	1430	241

SOURCE: Ministry of Labour

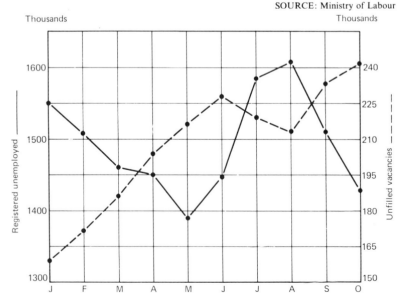

The general tendency in this case is for one graph to go up while the other goes down. Both these examples show that there may be some connection between the two variables.

Correlation

This connection between the sets of variables is known as correlation. Later the relation will be discussed further. A coefficient can be calculated to give a measure of the correlation, and if the variables simultaneously increase and decrease, the coefficient is positive, while if one increases as the other decreases, the coefficient is negative. So our first example could give positive or direct correlation, and the second negative or indirect correlation.

It is important to realise that a close correlation between two sets of figures does not necessarily mean that changes in the values of one variable are in any way caused by the changes in the other. There may be no connection between the changes, or both may be caused by a third event. This important fact is generally stated as 'correlation does not imply causation'.

Sectional Charts

When drawing bar charts, if an item consists of two or more parts, it is illustrated by a sectional bar chart. The same device is used in drawing graphs.

Example

Deliveries of Steel: Weekly Averages
(Thousand Tonnes)

	Qtr.	Home Deliveries	Exports
1975	1	477	86
	2	396	80
	3	336	63
	4	381	86
1976	1	435	84
	2	422	97
	3	360	93
	4	412	96
1977	1	411	101
	2	373	123
	3	358	119
	4	393	105
1978	1	463	100
	2	389	117
	3	341	102

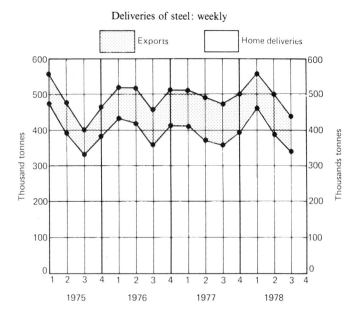

Deliveries of steel: weekly

Exports   Home deliveries

Thousand tonnes

Thousands tonnes

1975   1976   1977   1978

Exercise 7.1

1. Describe briefly the construction of statistical graphs.

2. State the use of statistical graphs.

3. What is meant by 'correlation'?

4. State when continuous variables are treated as discrete variables, and discrete variables treated as continuous variables. Give examples.

5. The following are the lighting-up times in London on certain days during the period of British Summer Time. From the graph estimate the lighting-up times on May 13th, June 24th, July 29th, August 26th, September 30th.

Lighting-up time	9·09	9·31	9·53	10·09	10·20
Date	Apr. 22	May 6	May 20	June 3	June 17
	10·20	10·11	9·41	9·15	8·44	8·12
	July 1	July 15	Aug. 5	Aug. 19	Sept. 2	Sept. 16

6. The following table gives the values of two variables tabulated over 12 months at intervals of one month. Draw in one diagram a graph of each of these variables on a time basis.

By comparing your graphs, state whether you consider there may be a connection between these variables. If so, indicate the nature of the connection, giving reasons for your opinion.

Interval	1	2	3	4	5	6	7	8	9	10	11	12
Variable A	3	12	15	2	5	20	10	6	25	18	18	27
Variable B	18	32	25	9	21	25	8	12	32	18	13	38

(L)

7. The following table gives the number of houses completed (in Hundreds) and the number of persons unemployed (in Ten thousands) for each month of the year 1947. Draw in one diagram, graphs of both series and state any connection between the graphs. Suggest a reason for any unusual feature in the diagram.

	Jan.	Feb.	Mar.	Apr.	May	June
Houses (Hundreds)	82	31	58	91	115	119
Unemployed (Ten thousands)	44	192	81	46	36	30
	July	Aug.	Sept.	Oct.	Nov.	Dec.
Houses (Hundreds)	115	116	112	98	96	91
Unemployed (Ten thousands)	28	27	26	28	29	31

(L)

8. Statistical data are frequently represented by diagrams or graphs. Describe the types of diagram or graph which you consider most suitable for the presentation of statistical data to the general public. Give reasons for your choice. (A.E.B.)

9. The following figures show consumer expenditure on wines and spirits in the U.K., prices being current. Graph the data, taking the horizontal axis to represent quarters and the vertical axis the expenditure. Write a paragraph describing the main features shown in your graph.

Consumer Expenditure on wines and spirits (£Million)

Quarters	1975	1976	1977	1978
1			507	829
2	414	545	568	1027
3	464	554	664	1069
4	565	697	789	

10.   Coal Lost Through Stoppages Arising from Certain Causes
(Thousand Tonnes)

	Holidays	Disputes	Accidents, repairs and breakdowns
Nov. 1947–Feb. 1948	1919	330	297
Mar. 1948–June 1948	2675	434	265
July 1948–Oct. 1948	4423	139	160
Nov. 1948–Feb. 1949	1893	228	138
Mar. 1949–June 1949	2572	561	147
July 1949–Oct. 1949	4606	529	120
Nov. 1949–Feb. 1950	1796	127	98
Mar. 1950–June 1950	2644	243	81
July 1950–Oct. 1950	4622	434	101
Nov. 1950–Feb. 1951	1766	176	86
Mar. 1951–June 1951	2629	363	142
July 1951–Oct. 1951	4663	265	71

Construct a single graph to illustrate the three time-series given in the table above, using a vertical scale of two cm to represent one thousand units. Describe, introducing rough material estimates where necessary, any trends, cycles or other conspicuous variational features of each series.   (*J.M.B.*)

11.   A graph is drawn to show the output of coal for each month of the year 1978. State the advantage and the disadvantage of using a scale on the vertical axis running from 140 to 220 million tonnes, given that in each month the output fell within this range.

(*L*)

12.   The disappearance of the farm horse in England and Wales.
Horses per 400 hectares of crops and grass

	Hertford	Westmorland	England and Wales
1901–03	34·1	23·8	34·1
1911–13	32·5	23·1	32·5
1921–23	27·6	25·1	31·1
1931–33	21·6	21·1	26·0
1939	15·8	20·6	22·3
1944	12·2	23·6	20·0
1947–48	8·4	20·2	16·2
1953	3·0	10·5	7·4
1958	1·1	4·2	2·9

Illustrate the above data by three graphs drawn on one diagram. Taking Hertford as typical of the counties near London, and Westmorland as typical of counties more distant, write comments on the disappearance of the farm horse in England and Wales.

(*J.M.B.*)

13. The table below shows the deaths, in thousands, from three diseases during a period of ten years:

Year	1947	1948	1949	1950	1951	1952	1953
Tuberculosis	23·9	22·9	17·5	14·1	12·0	9·3	7·9
Influenza	3·9	1·5	5·7	4·5	17·7	2·1	6·9
Pneumonia	26·2	20·7	23·6	20·3	25·6	21·0	23·0
Year	1954	1955	1956				
Tuberculosis	7·1	5·8	4·9				
Influenza	2·1	3·3	3·0				
Pneumonia	20·4	23·5	24·8				

Draw graphs of all three time series in one diagram. Write comments on the incidence of deaths from each disease.

(*J.M.B.*)

Ratio Scales

So far the graphs have been drawn with natural or arithmetic scales, that is scales which space out evenly the divisions corresponding to numbers increasing uniformly in size; that is, they show the actual or absolute change in the variable. Often it is more convenient to have a scale where *equal distances show equal percentage or relative changes*. This scale is known as a ratio or logarithmic scale, and depends on the properties of logarithms.

Construction of a Ratio Scale

To construct a ratio scale ranging from, say, 1 to 500 write down the numbers and their logarithms.

Number	Logarithm	Number	Logarithm	Number	Logarithm
1	0	60	1·778	100	2·000
10	1·000	70	1·845	200	2·301
20	1·301	80	1·903	300	2·477
30	1·477	90	1·954	400	2·602
40	1·602			500	2·699
50	1·699				

Then mark along the chosen axis distances proportionate to the logarithms and at these points mark the corresponding numbers as shown below.

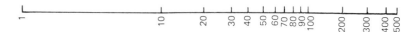

From 20 to 30 there is an increase of 50 per cent, which is represented on the scale by a distance proportionate to 0·176 (i.e. 1·477–1·301), whilst from 200 to 300 there is also an increase of 50 per cent, and the distance on the scale is again proportionate to 0·176 (2·477–2·301). So equal distances on a ratio scale always show equal percentage changes.

It should be noticed that there is no zero on a ratio scale. The logarithm of one is nought, and the logarithm of nought is minus infinity.

Use of Ratio Scale
Another advantage of a ratio scale is that large magnitudes can be represented within reasonable limits, as the range between the two extreme items is reduced. For example, the following table gives the wave-length in metres of some electro-magnetic waves.

Type of radiation	Wave-length in metres
Long Radio	$5·0 \times 10^2$
Short Radio	$1·0 \times 10^1$
Long Infra-red	$1·0 \times 10^{-4}$
Medium Visible Red	$8·0 \times 10^{-7}$
Medium Visible Violet	$4·1 \times 10^{-7}$
Long X-ray	$1·0 \times 10^{-7}$
Short X-ray	$3·0 \times 10^{-12}$
Long Cosmic Ray	$8·0 \times 10^{-14}$
Short Cosmic Ray	$3·0 \times 10^{-15}$

The longest wave is $1·67 \times 10^{17}$ times as great as the shortest wave, so if the shortest wave is represented by 0·1 cm on an ordinary scale, the longest wave is represented by $0·1 \times 1·67 \times 10^{17}$ cm, or $1·67 \times 10^{11}$ kilometres (i.e. over one thousand times the distance to the sun). But if these lengths are represented on a logarithmic scale, where say 0·5 cm represents a ratio of 10, the scale need be only 9 cm long. For:

$\log 5·0 \times 10^2 = 2·70;$ $\qquad \log 1·0 \times 10^1 = 1·00 = 1$
$\log 1·0 \times 10^{-4} = \bar{4}·00 = -4;$ $\qquad \log 8·0 \times 10^{-7} = \bar{7}·90 = -6·10;$
$\log 4·1 \times 10^{-7} = \bar{7}·61 = -6·39;$ $\qquad$ and so on.

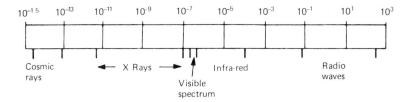

10^{-15}	10^{-13}	10^{-11}	10^{-9}	10^{-7}	10^{-5}	10^{-3}	10^{-1}	10^{1} 10^{3}

Cosmic rays   ← X Rays →   Visible spectrum   Infra-red   Radio waves

Logarithmic Graphs

If there are two variables and only one variable is represented on a logarithmic scale the graph is known as a semi-logarithmic graph, while if both variables are represented by logarithmic scales the graph is known as a logarithmic graph.

Semi-logarithmic graphs are useful to compare relative changes. Thus, if there is a boom in trade and a firm wishes to discover if they are experiencing the average relative rate of increase in their sales, they could plot on a semi-logarithmic paper the total sales per month for the industry in which they are interested and their own sales. If the lines are parallel it means their rate of progress is equal to the average.

Thus the following table gives a firm's annual sales and the annual sales of the industry as a whole:

Year	1972	1973	1974	1975	1976	1977	1978
Firm's Sales (£ Thousands)	1·2	1·6	1·9	2·6	3·0	4·8	5·6
Industry's Sales (£M)	3·4	5·0	5·9	8·2	11·0	14·7	15·1

Plotting these values on semi-logarithmic paper we have:

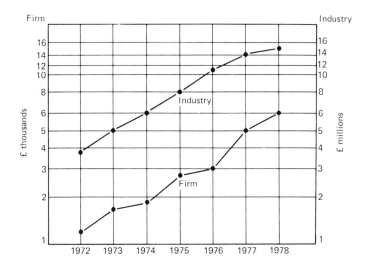

and since the lines are approximately parallel the firm is experiencing the average percentage increase in its sales.

Another example to illustrate the advantage of a ratio scale to represent the relative rate of increase in the cost of a commodity is as follows. Suppose an article cost £10 in 1950, £20 in 1960 and £35 in 1970. If the costs are plotted on a graph with a natural scale we get the impression that the cost of the article increased more rapidly between 1960 and 1970 than between 1950 and 1960. This is not so. Between 1950 and 1960 the cost of the article doubled while the increase between 1960 and 1970 was 75%. The explanation is that equal distances on the vertical scale of the chart represent equal absolute amounts. Such a chart is not suited to represent relative changes, which is what we want when examining rates of change.

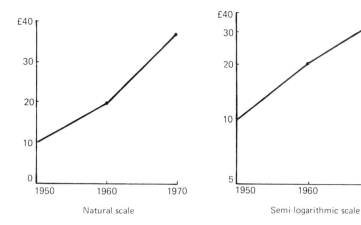

Natural scale   Semi-logarithmic scale

If we represent the cost of the article on a semi-logarithmic or ratio graph, equal percentage changes in the cost of the article are represented by equal vertical distances, and the fact that the rate of increase in the cost declined after 1960 is clearly shown.

When drawing the histogram for Agricultural Holdings (page 77) we had difficulty in representing the distribution effectively; there was a long thin column one end and very flat tail the other, and the detail of this tail was not observable. In such cases the heights of the columns can be plotted on a logarithmic scale.

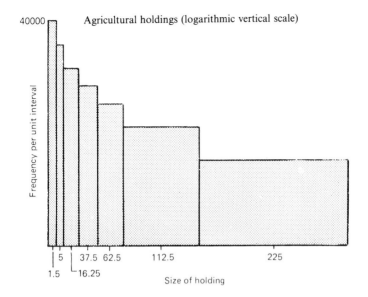

Agricultural holdings (logarithmic vertical scale)

Frequency per unit interval

Size of holding

If the range of values of the second variable is also large, this variable can also be plotted on a logarithmic scale. The shape of the histogram then becomes,

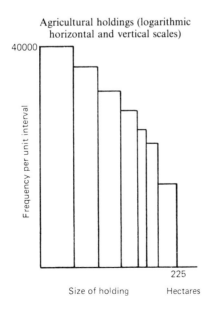

Agricultural holdings (logarithmic horizontal and vertical scales)

Frequency per unit interval

Size of holding   Hectares

Warning

It must be emphasised that histograms and graphs drawn with logarithmic scales should not be shown to persons not skilled in their interpretation. The shapes of the histograms above differ greatly from the shape of the original histogram and from each other, and an uninformed person could easily make mistakes in interpretation.

To the mathematician, the shape of a logarithmic graph can suggest a relation between the two quantities. Thus suppose the two quantities are x and y and the relation between them is represented on a logarithmic graph by a straight line. Then

$$\log y = k \log x + \text{const}$$
$$= k \log x + \log c \quad \text{say,}$$

Therefore $y = cx^k$.

Exercise 7.2

1. The following are the estimated times of events on the earth. If, on a natural scale, 1 cm represents 1 million years, calculate in m the lengths required to represent the other events.
    Represent on a logarithmic scale

	Million years		Million years
Age of earth	2000	Giant reptiles appear	250
First shell fossils appear	500	Giant reptiles disappear	100
First fishes appear	350	Present day mammals	
First plants appear	300	appear	50
First amphibians appear	275	Man appears	1

2. The following table gives the approximate distances from the earth of some astronomical bodies. If, on a natural scale, the distance of the sun is represented by 1 cm, calculate the distances required to represent the distances of the other objects.

	Kilometres		Kilometres
Distance of Sun	$1 \cdot 45 \times 10^8$	Galaxy in Hercules	$1 \cdot 3 \times 10^{21}$
Diameter of Solar		Galaxy in Gemini	$3 \cdot 1 \times 10^{21}$
System	$6 \cdot 4 \times 10^9$	Galaxy in Cygnus	$4 \cdot 8 \times 10^{21}$
Nearest Star	$4 \cdot 2 \times 10^{13}$	Galaxy in Bootes	$4 \cdot 3 \times 10^{22}$
Diameter of Galaxy	$9 \cdot 7 \times 10^{17}$	Limit of vision with	
Thickness of Galaxy	$1 \cdot 6 \times 10^{17}$	optical telescope	$4 \cdot 8 \times 10^{22}$
Magellanic Clouds	$1 \cdot 6 \times 10^{18}$	Limit of detection	
Andromeda	$1 \cdot 8 \times 10^{19}$	with radio telescope	$9 \cdot 7 \times 10^{22}$
		Estimated radius of Einstein Universe	$8 \cdot 0 \times 10^{23}$

3. Show, on a logarithmic graph, the distribution of personal incomes before tax for 1973/4.

Income £	No. of Incomes (Thousands)
595 under 750	1053
750 under 1000	1958
1000 under 1250	2000
1250 under 1500	1964
1500 under 1750	2031
1750 under 2000	1954
2000 under 2500	3670
2500 under 3000	2602
3000 under 4000	2299
4000 under 5000	663
5000 under 6000	227
6000 under 8000	202
8000 under 10 000	83
10 000 under 20 000	114
20 000 and over	21·6

SOURCE: Board of Inland Revenue

4. Show, on a logarithmic graph, the number and capital value of estates in Great Britain in 1965/66.

Net Value (£ Thousands)	Number	Net Value (£ Thousands)	Number
Less than 1	101 737	25–	7354
1–	48 149	50–	2702
2–	33 621	100–	1188
3–	41 825	500–	34
5–	34 560	1000–	14
10–	20 371		(A.)

5. Plot the population census figures for England and Wales given on page 89 on semi-logarithmic graph paper. From your graph, decide when the relative rate of increase of the population was (i) greatest, (ii) least.

6. Explain what is meant by a semi-logarithmic graph, and state briefly when you would use it in preference to a natural scale graph.

The table gives a firm's annual sales and the annual sales of the industry as a whole. Graph the two series on the same semi-logarithmic diagram so as to bring out their relative movements.

State briefly your conclusions.

	1968	1969	1970	1971	1972	1973	1974
Firm (£ Thousands)	1·6	1·8	1·95	2·15	2·35	2·6	3·6
Industry (£ Million)	3·1	3·8	4·7	5·7	7·0	8·4	10·5

	1975	1976	1977
	5	7	10
	12·5	15·5	19·0

(*A.E.B.*)

7. The Annual Report of the Ministry of Health shows the following figures.

Year	1967	1968	1969	1970	1971
No. of Prescriptions (Million)	271·2	267·4	264·2	266·6	266·5
Average Cost (p)	53·9	56·7	61·8	67·5	75·8

	1972	1973	1974	1975
	275·9	284·2	295·4	303·5
	82·4	88·3	100·7	127·9

SOURCE: Dept. of Health and Social Security

(*a*) Draw a semi-logarithmic chart to bring out the relationships in the above figures.

(*b*) Name four ways in which a semi-logarithmic chart differs from an arithmetic scale graph.

8 Collection of Statistical Data

Statistical Data

So far, the collection, tabulation, and diagrammatic presentation of statistical data have been discussed, but no mention has been made of the methods of obtaining information. The Government publishes a large amount of statistics on many aspects of the life of the nation; the United Nations on international affairs; private and nationalised industries on their own affairs; and scientific journals on scientific matters. Often the answer to an enquiry may be found amongst this vast amount of published material but the probability is that it will not be classified and presented in a way suitable for a problem under discussion. Thus a new enquiry may have to be undertaken, and, obviously the value of any result and the truth of its conclusions will depend upon the care and accuracy with which the enquiry is conducted. Furthermore it is a waste of time to apply refined statistical methods to material of doubtful accuracy.

Method of conducting an Enquiry

If an enquiry has to be undertaken, much time can be saved, and more accuracy obtained, by careful and detailed planning. An assessment of the problem should be made by:

1. *deciding precisely the nature of the problem.* It is advisable to express the problem on paper, and to keep on restating it until its precise nature is clearly evident. If the problem remains complex, it may be possible to separate it into two or more problems, and hold two or more investigations. The success of any enquiry depends on this step.

2. *listing any factors that may affect the investigation.* It is impossible to give an exhaustive list of factors, but it should include such factors as items given in different units, or different degrees of accuracy, the reluctance on the part of some persons to answer questions, the times at which the information is collected, ambiguity in the definition of terms, lack of instructions, the form of the final classification, and so on. The consideration of these, and any other factors that may affect the final analysis of the results is very

important. As an example, suppose an investigation is to be held into the heights of the boys in school. It is quite conceivable that, in answer to the request 'state your height', six boys of exactly the same height could give the following different answers, (*a*) about 152 cm, (*b*) 152 cm, (*c*) 1 m 52 cm, (*d*) 1 m 51·5 cm, (*e*) 151·5 cm, (*f*) 1·52 m. The diversity of these answers is due to no mention being made in the request of the unit to be employed and of the degree of accuracy expected, and this would involve the investigator in much unnecessay work that could have been avoided if thought had first been given to his requirements. A better request is—'State your height in cm, correct to the nearest cm. But, even in this case, error may arise in the classification due to the absence of instruction to the boy whose height is measured as 152·5 cm. A note should be added by the investigator stating in which class he wishes the boy to place himself: 152 cm or 153 cm. Time is well spent considering any contingency that may arise and affect the investigation.

3. *considering the possible methods of obtaining the information.* That is, whether to use data collected for a previous investigation, and which is then known as secondary data, or hold a new investigation and obtain primary data, or a combination of both; and if a new investigation is decided upon, whether it is necessary to hold a complete investigation of every item (known as the universe of the investigation), or whether a partial survey of the items will suffice; that is, by investigating a 'sample'. Sampling plays an important part in statistics, and will be discussed in detail later.

4. *the formation of a plan.* Any plan should be as complete as possible, but should not be too rigid to prevent those adjustments that will become necessary when the inevitable unexpected factor affects the investigation.

Example

As an example, consider the following:

The entrance to Y-shire School faces a main road, and it is suggested that it is unsafe to cross the road at this point owing to the large volume of traffic. The Headmaster requests Mr. Chips, the statistics teacher, to investigate the truth of this statement. Mr. Chips invites the fourth-form statistics group to help him, and the following results:

ASSESSMENT OF THE PROBLEM

1. *Object*

To investigate the volume of traffic at a point on the main road, known as North–South, near the entrance to Y-shire School.

2. *Factors affecting the investigation*
   (*a*) Most pupils use the road between 8.30 a.m. and 9.0 a.m., 12.30 p.m. and 1.30 p.m., 4.0 p.m. and 4.30 p.m.
   (*b*) Market-day is on Saturday, when the school is closed.
   (*c*) Early closing day is Thursday.
   (*d*) Classification. A pedal-cycle is not so dangerous as a fast-moving car. Therefore it is important to know the number of each type of vehicle.
   (*e*) The direction of the traffic—whether there are equal streams of traffic in each direction or one dense stream in one direction.
   (*f*) A census of traffic was taken at this spot five years ago. The figures giving the density of traffic per hour are still available.
   (*g*) A new industrial business opened on the outskirts of the town two years ago.

3. *Courses*
   (*a*) Use the figure of the previous census.
   (*b*) Conduct a new census.

4. *Conclusion*
   The opening of the industrial business, and the general increase in the volume of traffic have made the results of the previous census unreliable. Therefore a new census will be taken. Wednesday is a typical day for the purpose of the census.

PLAN

Object
To conduct a census of traffic on the North–South Road near the entrance to Y-shire School.

Information
The Headmaster requires a census of the traffic passing the school to estimate the danger to pupils in crossing the road at the entrance to the school.

Intention
The census will be taken on Wednesday, (Date), for the following periods: 8.30 a.m. to 9.0 a.m., 12.30 p.m. to 1.30 p.m., 4.0 p.m. to 4.30 p.m.

Method
1. The traffic will be classified as (1) Horse-drawn, (2) Pedal-cycles, (3) Motor-cycles, (4) Motor-cars, (5) Vans, (6) Lorries.

2. Two recorders will form a team, and one team will be stationed at each side of the road for each period.
3. The two recorders on the west side will count the traffic going north, and the two recorders on the east side will count the traffic going south.
4. In each team the first named recorder will count the horse-drawn vehicles, pedal-cycles and motor-cycles, and the second recorder the motor-cars, vans and lorries.
5. To assist in counting, recorders will record each vehicles by a stroke in the appropriate space on the recording sheet provided, each fifth stroke being made diagonally through the previous four—thus ⅲⅢ.
6. The following is the duty rota.

Time	West Side of Road	East Side of Road
8.30 a.m.–9.0 a.m.	H. Wharton, J. Brown	R. Cherry, D. Miller
12.30 p.m.–1.30 p.m.	J. Merryman, D. Dare	J. Halifax, E. Little
4.0 p.m.–4.30 p.m.	H. Weed, S. Nelson	J. Arnold, S. Luck

7. Fourteen copies of the recorder's sheet will be required.
8. At the end of each duty period the recorders will immediately take their sheets to Mr. Chips.

COPY OF THE RECORDER'S SHEET
TRAFFIC CENSUS

North-South Road at the entrance to Y-shire School.
Name of Recorder .... Date................
Position of Recorder .... Period of Duty.......

	Horse Drawn[1]	Pedal Cycle[2]	Motor Cycle[3]	Motor Cars[4]	Vans[5]	Lorries[6]
Totals						

Notes: [1] Any horse-drawn vehicles.
[2] Any non-motorised bicycle, tricycle, or invalid carriage.
[3] All motorised bicycles, tricycles, or invalid carriages.

(4) All types of motor-cars normally used for transportation of persons (see note 6 re buses). A motor-car with a trailer or caravan attached to be counted as two vehicles.

(5) All types of closed motor vehicles used for transporting light goods, or furniture, including electrically propelled milk and bread delivery vans. Ambulances.

(6) Any open vehicle used for delivering heavy loads, fire engines, locomotive ploughing engines, tractors, agricultural tractors, trench diggers, mobile cranes, and similar motorised mechanical implements. Tankers used for transporting petrol, milk, chemicals in bulk. Dust-carts. Road rollers. Buses used for public transport (whether municipally or privately owned).

Remarks

No date has been placed on the sheet in case the census has to be postponed. A completely different plan could be drawn up, but the decisions taken must be left to the person in charge of the investigation, who should be the one with a clear idea of the information required, and the local environment.

The important part of the example is the routine to be followed. First the assessment consisting of (1) object, (2) factors affecting the investigation, (3) courses open to the investigator, (4) conclusion. Second the detailed plan, consisting of (1) object, (2) relevant information, (3) intention, (4) method.

Person helping in an investigation will only take the necessary care and interest if they are fully convinced that the purpose of the investigation is important, and that adequate care and thought have been taken to make their duties clear.

9 Questionnaires

Questionnaires

The information required for the example in the last chapter was obtained without the necessity of relying upon the goodwill of others to answer questions, but many investigations will depend upon this goodwill. The form listing the questions is known as a *questionnaire*, and the drawing up of a questionnaire requires skill and practice.

Many government departments legally demand that a person supplies the answers to certain questions, and threaten the person with dire consequences if he fails to do so, or wilfully supplies the wrong information. But, as previously stated, most investigators depend only on the goodwill of the person questioned to supply the answers, and to achieve this, considerable thought must be given to the form of the questionnaire. It must clearly state who requires the information and for what purpose it is required, in order to allay any suspicion on the part of the person questioned.

Drawing up of a Questionnaire

Before proceding with a questionnaire, the investigator will have drawn up an assessment of the problem, and will have written down a precise statement of the information required. Thus his ideas should be clearly defined and this should make formulation of questions easier. Still, first attempts will probably be unsatisfactory, one question being too vague; another requiring excessive trouble to answer truthfully; another may necessitate the insertion of an extra one to act as a check on the answer; the wording of another may tend to influence the answer; while another may arouse the opposition of the person questioned.

Always be guided by the following rules:

The questions should be

1. courteously phrased.
2. clearly phrased.
3. few in number.
4. capable of being answered without undue inconvenience.

5. capable of being answered by the least intelligent and least educated person questioned.
6. preferably capable of being answered in a definite manner, that is, by a yes or a no, a number, a place, a tick, a date, and so on.
7. capable of being answered truthfully and willingly.
8. not capable of being judged as prying.

Even if the above rules are scrupulously obeyed, the probability is that some of the replies will still be worthless, and will have to be discarded. The degree of success will depend to a large extent upon the care and thought given to the draft questionnaire.

Example
As an example, suppose the Statistics Group of Y-shire School decided to investigate the number of times the pupils of the school visited the cinema. The following would be a suitable questionnaire.

Y-shire School Statistics Group

Date . . . .

For the purpose of an exercise in statistics the pupils of the above group would be pleased if you would complete the questionnaire set out below concerning your visits to the cinema during the last year.

It is stressed that the answers are required for an exercise in statistics and for no other reason.

Your co-operation will be greatly appreciated.

. . . .

Number of Visits to the Cinema During the Last Year

Name . . . .   Form
Age in years and months. . . . .
How often have you been to the cinema in the last year?
Place a tick on the line after the most correct answer.

(*a*) None . . . .   (*f*) Once a week . . . .
(*b*) Only a few times . . . .   (*g*) Twice a week . . . .
(*c*) Once a month. . . . .   (*h*) Three times a week . . . .
(*d*) Twice a month . . . .   (*i*) Four times a week. . . . .
(*e*) Three times a month . . . . . .   (*j*) Five or more times a week. . .

111

Exercise 9.1

1. You are requested to conduct the following inquiries. Draw up assessments, plan, and questionnaires to find out from the pupils in the school:
   (a) the time spent viewing television.
   (b) which daily paper is read.
   (c) which form of school sport is preferred.
   (d) the mode of transport to and from school.
   (e) the amount of time and care given to bicycles.

2. What rules should be observed in compiling questionnaires to ensure a successful investigation? (L)

3. Write a brief account of any statistical investigation in which you have assisted, giving details of:
   (a) the sources from which the data were collected;
   (b) the methods by which the data were collected;
   (c) the object of the investigation and any conclusions which were reached. (L)

4. State briefly what you consider to be the essentials of a good questionnaire, referring particularly to the type of question to be asked and the kind of answer required.
   It is desired to set up in your district a club for young people. Draw up a questionnaire of from six to ten questions designed to help a committee to decide the form the club should take, the times and places of meetings and other details considered necessary to ensure the success of the club. (A.E.B.)

5. Your form is asked to investigate the amount of money spent per week by the pupils of your school on travelling to and from school. Draft a suitable questionnaire. (A.E.B.)

6. A group of pupils wished to investigate whether there was correlation between the age and height of the pupils in their school. To obtain the necessary information, they gave each pupil a copy of the following questionnaire:

ABC School
Statistical Investigation, 1978
State your (i)   Name................
          (ii) Age ....
          (iii) Height ....

112

(i) List the faults of the above questionnaire.

(ii) Rewrite the questionnaire in a form you consider suitable.

(A.E.B.)

7. List a set of rules as a guide to the formulation of questions for a questionnaire. *(A.E.B.)*

10 Sampling and Bias

Sampling

When carrying out an investigation, the investigator may find that, at times, it is impossible to examine every item with the relevant characteristic (known as the parent universe). In this case all he can do is to examine a limited number of items, and hope that from this limited number he can obtain the information he requires about the parent universe. The group of items selected for the investigation is known as a 'sample', and the act of investigating this selected group as 'sampling'. A sample can be of any size, and may consist of the entire universe. Sampling will be necessary if the parent universe is so large that a complete investigation would be prohibitive because of the expense involved in time or money; or because the act of investigating may destroy the article; for example, in finding the life span of a pair of shoes, or in testing an apple for sweetness.

Properties of a Sample

To investigate the properties of a sample, the following experiment was conducted. Twelve red and twelve black cards were taken from a pack of playing cards and thoroughly shuffled. A sample of either 1, 2, 3, . . . or 24 cards was taken from these and the number of black cards in the group of twenty-four cards was estimated to the nearest whole number. After each selection the sample was replaced and the cards thoroughly shuffled before taking another sample. This was repeated fifty times for each size of sample. The estimated number of black cards in the set of twenty-four cards was tabulated against the size of the sample. The results on p. 115 were obtained.

Take, for example, the column relating to sample size of 9 cards. The table shows how the 50 samples came out and the calculations required to estimate the number of black cards in the universe in each case.

Result of Sampling

Estimated Number of Black Cards	1	2	3	4	5	6	7	8	9	10	11	12	13	14	15	16	17	18	19	20	21	22	23	24
0	22	15	8	2	1	2	1																	
1																								
2																								
3							3																	
4					5	5																		
5				13							4													
6								6	7	5			5											
7			20			10				4			4											
8		21			20				8		8	9				4	4	3						
9				21			14			11		10	5	7	6		7	4	3					
10						18		17			12				15	9	19	11		7	12	12	24	50
11					17						14	14	14	20		17				25	14	25	26	
12							18		15						11			20	22			13		
13		21		8		8		11		12	8	8	9	6		14	14	9	22	15	16			
14			10						9			9		7		6	6		3	3	8			
15							7	5					6		6			3						
16					6								4	5	6									
17							2																	
18						6				4			3	5										
19											4													
20																								
21																								
22																								
23																								
24	28	14	12	6	1	1																		

Size of Sample

115

No. of black cards in sample	No. of times found	No. in universe	Nearest integer
2	7	$\frac{2}{9} \times 24 = 5\cdot3$	5
3	8	$\frac{3}{9} \times 24 = 8$	8
4	11	$\frac{4}{9} \times 24 = 10\cdot7$	11
5	15	$\frac{5}{9} \times 24 = 13\cdot3$	13
6	9	$\frac{6}{9} \times 24 = 16$	16
	50		

This investigation illustrates that:

1. Any sample will give some information about the parent universe. A sample of one card tells us that there are either black or red cards.
2. As long as the sample is large enough to avoid the influence of abnormalities it will be representative of the characteristics of the parent universe. This is the law of statistical regularity.
3. The larger the sample becomes, the more accurate is the estimation.
4. The larger samples show a higher degree of stability than the smaller samples; that is, the estimate for the number of black cards remains steadier for the larger samples.
5. Even if a sample is chosen correctly it will not necessarily be exactly representative of the parent universe, for all samples are not alike. Thus there is always a sampling error, but its size decreases as the size of the sample increases.

Methods of Sampling

The purpose of sampling is to obtain information about a particular characteristic or characteristics of the parent universe, and it is important to be clear on the information required. This will decide the sampling method to be adopted.

Sampling may be carried out by selecting the required number of items:

1. at random
2. by some purposive principle
3. by a mixture of 1 and 2.

Sampling of type 1 is known as random sampling; of type 2 as purposive sampling; and of type 3 as mixed sampling.

If the universe is divided into 'strata' by purposive methods, the sample is said to be stratified.

Random Sampling

If each member of the universe has the same chance of being selected, the selection is then random. The selection of a random sample is more difficult than it at first appears. The human being is an extremely poor instrument for making a random selection, and will invariably make a personal choice.

Any one-sided influence that affects a selection is known as bias.

The methods adopted for random sampling depend upon the size and nature of the universe. If there is only a small number of items, each item can be given a number and corresponding numbers can be placed on cards. The cards can be well shuffled, and the number corresponding to the size of the sample dealt from the pack. The items with numbers corresponding to those on the cards dealt would be the items forming the sample. If the universe is large, this is an impracticable method, and it is better to select random numbers from a table of random sampling numbers. This is a table of numbers constructed with great care to ensure their randomness. The following is part of a table of random numbers:

3063 7752 4300 0803 8080 6022 4280 6735 5278 0635 2656 2407 8314
4676 2411 3704 3481 8594 4519 3873 4674 2190 7024 8702 1671 6357
2223 7218 7119 9237 5344 9887 3865 0254 2516 0136 8113 7222 9927
7557

The table is used as follows: suppose ten numbers from 0 to 60 are required. We start anywhere in the table and write down the numbers in pairs. The table can be read horizontally, vertically, diagonally, or in any *methodical* way. Starting at the first and reading horizontally, we obtain 30, 63, 77, 52, 43, 00, 08, 03, 80, 80, 60, 22, 42, 80, 67, 35, 52, and so on. Ignoring the numbers greater than 60, we obtain for our ten random numbers 30, 52, 43, 0, 8, 3, 60, 22, 42, 35.

Random sampling may produce the most unrandom-looking results. For example, thirteen cards from a well shuffled pack of playing cards may consist of one suit. Also we can never be certain that a method of sampling is random. There are always conceivable sources of bias and we must take care to try to eliminate them. A method suitable for the selection of a random sample from one population need not necessarily be the best method to use when sampling another population.

Purposive Sampling

The danger of a purposive sample is the introduction of bias. Suppose a teacher was asked to select three boys from his form as a sample. He would probably select what he considers to be 'average' boys, that is average in behaviour, cleanliness, work, and any other quality he may have in mind. This may give us a sample that is typical or representative of the form, while a random sample may give us three boys who differ very widely from the average.

The purposive sample when small has the advantage of being typical, but if it is made larger, it remains an average sample. However a random sample, if made larger, will become more and more representative of the form, including the extremes, and the purpose of a sample is to give us information about the whole universe.

Stratified Sampling
Stratified sampling is a combination of random and purposive sampling, and to a large extent avoids the possible abnormalities of random sampling, and the bias of purposive sampling.

Thus, if we wish to ascertain the opinion of boys in a school on a certain matter, and we proceed to select a sample by random methods, we might get a sample consisting of sixth form boys only. To avoid this, we purposely select boys from each 'stratum' in the school, i.e. from each form. The boys from each form could be selected by a random method.

No rules can be laid down for sampling. The methods used must depend on the circumstances of the case, and unless the methods are random the reliability of the results is a matter of opinion.

Errors Due to Bias
It must be stressed that errors due to bias can render useless the results of an investigation. These errors differ from random sampling errors in that they increase with the size of the sample and their effect cannot be calculated.

Bias is always introduced if there is:
1. Any form of deliberate selection, as mentioned in the paragraph on purposive sampling.
2. Substitution of another item for one chosen in a random sample. Thus if it were decided to interview every tenth householder in a street and one was not at home, it would be wrong to interview the eleventh or any other number in his place, as the characteristics of persons who remain at home for a large part of the day differ from those who spend a large part of their time away from home.
3. Omission of part of a sample. Thus, if a random sample is chosen and a questionnaire is sent by post, unless every member of the sample replies, the result will be biased, as only those who have time for form filling, or who wish to influence the result of the investigation, will reply.

4. Replacement of a random selection by a casual or haphazard selection.
5. An appeal to the vanity of the person questioned. For example, if the question 'Are you a good athlete?' was posed, probably most people would succumb to vanity and answer 'Yes'.
6. An appeal to sympathy. Often an interviewer must make himself pleasant to the person questioned to obtain an answer, and if he is too successful, the person questioned will tend to give the answer he imagines the interviewer requires. This is particularly so with children.

This list indicates some possible causes of the introduction of bias, and is certainly not complete.

Exercise 10.1

1. How would you arrange to take an unbiased sample of
   (a) the inhabitants in a street?
   (b) the opinions of the boys in school on their favourite school game?
   (c) potatoes in a field?
   (d) the milk delivered at school?
2. A newspaper editor wishes to form an opinion of the public's reaction to a matter discussed in Parliament, and invites readers of his paper to write to him expressing their views. Is this a good sample?
3. An investigator wishes to know the opinion of householders on a certain matter, and issues the following instructions to his assistants:
   Visit every tenth house in each street in the district allocated to you. If there is no reply, call at the eleventh house and so on. Is this sample biased?
4. Make a random selection of ten boys from the form. Find the average height of the sample, and compare it with the average height of the boys in the form.
5. Ask a large number of persons to make a random choice of one of the following colours—red, orange, yellow, green, blue. Draw a bar chart of the result. Is there any bias?
6. Repeat the above investigation substituting the numbers 0, 1, 2, 3, 4, 5, 6, 7, 8, 9.
7. From strips of paper cut about 300 pieces between 0 and 10 cm long. Using a plastic ruler to avoid parallax, and using the side marked in millimetres, estimate the length of each piece of paper

to the one-hundredth of a cm. Draw a histogram of the frequency distribution of the last figure in the measurements. What do you conclude?

8. Use the table of random sampling numbers to select 60 numbers from 1 to 80. Find the average of the sample, and compare it with the average of the numbers 1 to 80, i.e. 40·5.

   Repeat for samples of 50, 40, 30, 20, 10, numbers.

   What conclusions do you draw?

9. Using a published set of winning numbers of Premium Bonds, test whether 'Ernie' (Electronic Random Number Indicating Equipment) makes a random selection of the numbers 0 to 9.

   (If the selection is random, each digit should appear approximately the same number of times.)

10. List six emotions and idiosyncrasies to which people are prone, and suggest circumstances or questions that would cause these characteristics to make people give biased answers to an interviewer.

    Suggest methods to avoid the possibility of bias.

11. What do you understand by 'sampling'?

    Describe briefly the three following methods:

    (a) random choice;

    (b) division into classes;

    (c) personal selection.

    Indicate in each case the precautions to be observed to avoid bias. (L)

12. 'It is usually impossible to measure all the values of any variable, the data available being only a sample of the total possible observations.' Illustrate this by reference to any statistical investigation in which you have assisted, giving details of the methods employed and indicating any conclusions reached. (L)

13. What do you understand by 'sampling'?

    In order to determine a new cost of living index, it is proposed to make a survey of the income and expenditure of 1000 households in a large city. Describe carefully two methods which might be used to select the sample households. (L)

14. Explain briefly what is meant by the term 'sampling'.

    A firm making mass-produced articles found that 7·9 per thousand were faulty and had to be rejected as unfit for sale. In 1955 the firm estimated that it has made a total of 72 000 articles (to the nearest thousand). Calculate the number which were fit for sale.

    Comment briefly on the accuracy of your answer. (L)

15. It is required to obtain a representative sample of 1000 people for an investigation into reading habits. Comment on the following methods for obtaining the sample:
    (a) By choosing 1000 names from the telephone directory.
    (b) By stopping 1000 people at random outside a main line station.
    (c) By asking 100 librarians to supply 10 names each.   (L)
16. Write an account of the methods employed to collect statistical data in (a) large surveys such as a population census, (b) small surveys such as might be carried out in a school or classroom. In the case of the small surveys explain how you would deal with bias.
                                                      (A.E.B.)
17. An interviewer is carrying out a short survey on a Friday afternoon. He walks about a busy shopping-centre selecting people at random and asks each person for an estimate of his or her weekly expenditure on alcoholic drinks. He then uses the mean of these estimates as an estimate of the average person's expenditure on alcoholic drinks.

    Is his sample likely to be an unbiased sample of the general population? Is the information given in reponse to his question likely to be free from bias? Examine in the light of these considerations the sources and the nature of such bias in his estimate as is likely to occur.   (J.M.B.)
18. Discuss briefly any bias which is likely to arise in response to the following questions, when directed by an interviewer at members of the class of people indicated.
    (a) Do you sympathise with Communism? (general public).
    (b) Do you use 'Surge' (a detergent)? (housewives).
    (c) What is your age? (general public).   (J.M.B.)
19. In a survey designed to ascertain the views of grammar school pupils, the following questions are put to samples of boys from grammar schools, the samples being selected by the headmasters:
    (a) Do you think that the more modern subjects, such as Physics, Chemistry, and Biology, are more important than the traditional subjects, such as History, Latin, and Greek?
    (b) Do you intend to stay on at school after the minimum leaving age permitted by law?
    Criticise the questions and the method of selecting the sample from the point of view of potential bias.   (J.M.B.)
20. Define the terms (i) bias, (ii) random sampling.
    Explain how errors introduced by bias differ from errors introduced by random sampling.   (A.E.B.)

21. (a) What is meant by (i) sampling, (ii) bias?

It is required to obtain a sample of 100 pupils from a school for an investigation into compulsory games. State how you would select your sample so that it is representative of the whole school, and the precautions you would take to avoid bias.

(b) Comment briefly on the statistical inference in the following statement.

'There are about 500 000 teenage pupils in grammar schools in England and Wales. We sent questionnaires to the Head Teachers of these schools and received replies from over 50%. Of their teenage pupils we learnt that only 0·1% had been in any kind of trouble with the police. As our sample is large we conclude that the modern teenager is a far better individual than we are often led to believe.'   (*A.E.B.*)

22. Explain what is meant by a random sample.

Describe the type of question suitable for a sample survey.

Construct a questionnaire for use in a school to investigate the television-viewing habits of the pupils.   (*L*)

23. An investigation into public opinion on purchase tax is to be made by
  (i)   interviewing people in the street,
  (ii) asking questions by telephone,
  (iii) sending a letter to every householder.

Criticise briefly each method.   (*L*)

11 Averages

When it is necessary to make comparisons between groups of numbers it is convenient to have one figure that is representative of each group. This figure is called the average of the group.

Thus at the end of a cricket season the performances of players are abbreviated and given as 'batting averages', and 'bowling averages', and these representative figures are used to compare individual performances.

There are many averages in use, and the choice of average depends upon which best represents the property of the group under discussion.

The Arithmetic Mean

This is the average of everyday use, and is defined as the sum of the items divided by the number of items.

Thus the arithmetic mean (A.M.) of 4, 7, 8, 13, is

$$\frac{4+7+8+13}{4} = 8$$

The calculation of an A.M. can often be simplified.

Example
Find the A.M. of 1936·4, 1902·4, 1921·4, 1906·5.

The sum is $(1900+36·4)+(1900+2·4)+(1900+21·4)+(1900+6·5)$
$$= 4 \times 1900 + 66·7$$

Therefore, A.M. $\dfrac{4 \times 1900 + 66·7}{4} = 1900 + \dfrac{66·7}{4}$

$$= 1900 + 16·7 = 1916·7.$$

1900 is called the *working mean* (W.M.), and 36·4, 2·4, 21·4, and 6·5, the *deviations from the working mean*. The deviations can be negative as well as positive. Any value can be chosen for the working mean, but the nearer the value is to the A.M. the easier the calculation becomes.

Example

Find the A.M. of 834·2, 788·1, 804·3, 760·2.

Take 800 as a working mean.

(Then 834·2 = 800 + 34·2,   788·1 = 800 − 11·9,

804·3 = 800 + 4·3,   760·2 = 800 − 39·8.)

Therefore A.M. $= 800 + \dfrac{34·2 - 11·9 + 4·3 - 39·8}{4}$

$= 800 - 3·3 = 796·7.$

Note and learn: *The arithmetic mean equals the working mean plus the average of the deviations from the working mean.*

Further Simplification

Example

Find the A.M. of 363, 484, 605, 726, 847.

These numbers have a common factor of 121.

Therefore, the A.M. $= \dfrac{121(3 + 4 + 5 + 6 + 7)}{5}$

$= \dfrac{121 \times 25}{5} = 605.$

These two methods of simplification can often be combined.

Example

Find the A.M. of 1075, 1225, 1375, 775.

Take a working mean of 1000.

Then A.M. $= 1000 + \dfrac{75 + 225 + 375 - 225}{4}$

$= 1000 + \dfrac{75(1 + 3 + 5 - 3)}{4}$

$= 1000 + \dfrac{75 \times 6}{4}$

$= 1112·5.$

Properties of the Arithmetic Mean

The arithmetic mean is the average of everyday use. It is the one most often used because it is easily understood, makes use of all the observations, is representative of the whole distribution, and is easy to manipulate mathematically.

Exercise 11.1

1. Find the arithmetic mean of 754, 723, 763, 714, 790, by
    (*a*) adding and dividing by 5,
    (*b*) taking 700 as a working mean,
    (*c*) taking 750 as a working mean,
    (*d*) taking 800 as a working mean.

2. By choosing a suitable working mean find the arithmetic mean of
    (*a*) 3, 6, 10, 14, 17, 19, 23.
    (*b*) 321, 452, 334, 521, 401, 483.
    (*c*) 3·6, 2·9, 4·7, 3·1, 4·2, 4·0.
    (*d*) $33\frac{1}{2}$, $37\frac{1}{2}$, $29\frac{1}{2}$, $24\frac{1}{2}$.
    (*e*) 79, 11, 52, 133, 144, 121.
    (*f*) $2\frac{1}{4}$, $3\frac{1}{8}$, $4\frac{3}{4}$, $5\frac{1}{2}$, $6\frac{3}{4}$.
    (*g*) 441, 426, 417, 364, 372, 365, 385, 420, 441, 432.

3. Find the H.C.F. of the following numbers and use it to find their arithmetic mean.
    (*a*) 189, 882, 1071.
    (*b*) 84, 144, 264, 360, 420.
    (*c*) 105, 147, 231, 252, 294.
    (*d*) 333, 1221, 1443, 2442, 555.

The Median

If the items of a discrete distribution are *arranged in order of size* starting with either the largest or smallest then the arrangement is called an *array* and the median is the value of the middle term.

Example
Find the median of 45, 51, 47, 50, 47, 50, 48, 50, 49.
    Arranging in order of size starting with the smallest, we have 45, 47, 47, 48, 49, 50, 50, 50, 51.
    There are nine items, and the middle one is the fifth.
    Therefore the median is 49.
    If there is an even number of items, the median is the arithmetic mean of the two middle terms.

Example
Find the median of 30, 56, 31, 55, 43, 44.
    Arranging in order we have, 30, 31, 43, 44, 55, 56.
    There is an even number of terms. The two middle terms are 43 and 44.

Therefore the median is $\dfrac{43+44}{2} = 43\cdot5.$

The middle term of a discrete distribution of size n is the $\dfrac{n+1}{2}$th item.

This ensures that we arrive at the same item whether we arrange the items in ascending or descending order of magnitude. Thus, in the above example the median item is the $\dfrac{6+1}{2}$, that is, the $3\frac{1}{2}$th item, and this gives us the magnitude of the median as $\dfrac{43+44}{2}$ in both arrangements.

$$\underbrace{\left\lvert\; 30 \quad 31 \quad 43 \uparrow\right.}_{3\frac{1}{2}\ \text{items}} \qquad \underbrace{\left.\uparrow 44 \quad 55 \quad 56 \;\right\rvert}_{3\frac{1}{2}\ \text{items}}$$

Properties of the Median

The median is useful if the exact values of extreme observations are unknown. For example it would be difficult to calculate an A.M. for the following distribution owing to the difficulty presented by both extreme classes, but the median can be calculated both definitely and easily.

Number of children attending school in a given area

Age	Under 5	5	6	7	8	9	10	11
Number	197	70	74	83	77	63	70	64

Age	12	13	14	15 and over
Number	60	54	55	227

There are 1094 children. Therefore the median is the arithmetic mean of the 547, and 548th items. Both these items lie in the class 9 years. Therefore the median class is 9 years.

Exercise 11.2

Find the median of the following sets of numbers, 1–5:
1. 41, 26, 17, 64, 72, 65, 85, 20, 41.
2. 32, 72, 99, 44, 57, 71.
3. 85, 78, 78, 94, 68, 10, 55, 31, 63, 87.
4. 88, 6, 96, 34, 28, 18, 28, 28, 70, 70, 70, 70, 56, 96.
5. 34, 28, 70, 56, 96, 21, 78, 22, 78, 87, 22, 78, 28.
6. The following table gives the ages and number of women who were married last year in a certain town. Calculate the median age.

126

Age	Under 21	21	22	23	24	25	26	Over 26
Number	12	11	14	17	26	25	16	57

The Mode

The mode of a discrete distribution is the most frequent, or most 'popular' item, and the modal class is that with the greatest frequency.

Example
Find the mode of the following distribution:

43, 27, 36, 27, 36, 29, 40, 36, 33.

43 occurs once.	29 occurs once.
27 occurs twice.	40 occurs once.
36 occurs three times.	33 occurs once.

36 occurs three times, and this is the most frequent item. Therefore the mode is 36.

Properties of the Mode

The mode is the most useful average in many problems. For example, if a new estate is being planned for an industrial firm it is important that the architect should know the most 'fashionable' number of children per family, so that he can build a number of houses to fit the needs of this size of family. The arithmetic mean of the number of children per family, say 2·5, does not give the necessary information. There may be half the number of families with no children, and half with five children.

Exercise 11.3

State the modal class in the following examples 1–5:

1. | Class | 1 | 2 | 3 | 4 | 5 | 6 | 7 | 8 | 9 | 10 |
   |---|---|---|---|---|---|---|---|---|---|---|
   | Frequency | 4 | 6 | 13 | 15 | 16 | 17 | 14 | 7 | 5 | 3 |

2. | Class | 1 | 3 | 5 | 7 | 9 | 11 |
   |---|---|---|---|---|---|---|
   | Frequency | 10 | 20 | 30 | 50 | 70 | 40 |

3. | Class | 1 | 11 | 21 | 31 | 41 |
   |---|---|---|---|---|---|
   | Frequency | 4 | 5 | 6 | 7 | 8 |

4. | Class | 10 | 20 | 30 | 40 | 50 |
   |---|---|---|---|---|---|
   | Frequency | 5 | 7 | 12 | 10 | 3 |

5. | Class | £5 | £7 | £9 | £11 | £13 |
   |---|---|---|---|---|---|
   | Frequency | 7 | 9 | 14 | 21 | 6 |

6. In breeding experiments with mice it was found that the size of the litter and the number of litters was,

Number in litter   1 2 3 4 5 6 7 8 9
Number of litters   6 10 13 15 25 32 12 1 1

State the mode of the distribution.

The Geometric Mean

The geometric mean (G.M.) is found by multiplying the n (say) items of the distribution together, and then finding the nth root of the product.

Example

Find the G.M. of 4, 7, 9, 13, 15.

There are five items. Therefore the

$$\text{G.M.} = \sqrt[5]{4 \times 7 \times 9 \times 13 \times 15} = 8 \cdot 7$$

Properties of the Geometric Mean

The advantage of this average over the A.M. is that it reduces the effect of a single large item on the average. It is used in problems related to growth such as the average size of population of an area over a given period.

Example

The A.M. of 1, 2, 3, 4, 4, 4, 5, 100, is 15·4, and this is not representative of the group in any way.

The G.M. is $\sqrt[8]{1 \times 2 \times 3 \times 4 \times 4 \times 4 \times 5 \times 100} = 4 \cdot 6$.

This is representative of the group.

Measures of Central Tendency

Since averages generally tend to lie centrally within distributions arranged in order of magnitude, they are sometimes called *measures of central tendency*.

Exercise 11.4

Find the G.M. of 1–4:

1. 4, 7, 9, 21, 100.   3. 63, 21, 42, 33, 201.
2. 10, 13, 41, 71.   4. 52, 27, 33, 45, 61, 900.

Exercise 11.5

1. Calculate the arithmetic mean of the following items:
   980·8, 981·1, 980·7, 980·3, 981·8, 982·5.   (L)

2. The dates of birth of a group of 10 girls are as follows:

Day   6   13 26   3 12 24   5 19 29   7
Month   5 12   3   7   6   8   1 11   3   2
Year   53 52 53 53 53 51 52 52 52 52

Tabulate the ages of the girls on 1st January, 1978, ignoring in each case fractions of one month. Calculate the average age of the group on January 1st, giving your answer correct to the nearest month.   (L)

3. The marks of 16 candidates in an examination were: 61, 53, 40, 46, 52, 50, 37, 41, 71, 38, 47, 64, 54, 45, 32, 73.

Find the average mark and the median mark.

Two further candidates took the examination later, and raised the average by 0·25 marks. Find the average mark of these two candidates.   (L)

4. The number of admissions to cinemas in Great Britain were:

1952	3rd	quarter	335·2 millions
	4th	,,	304·8
1953	1st	,,	328·5
	2nd	,,	323·5
	3rd	,,	326·7
	4th	,,	305·8
1954	1st	,,	325·7

Calculate the arithmetic mean of the number of admissions. In which quarters was the variation from this mean the greatest and least, and by how much?   (L)

5. Calculate the arithmetic mean of the following numbers: 390, 392, 399, 404, 391, 396, 387, 394, 386, 400, 395, 382.

If there is a possible error of ± 5 in the case of the even numbers and of ± 8 in the case of the odd numbers, find (a) the greatest (b) the least possible arithmetic mean.   (L)

6. The following figures give the number of children killed each month from April 1957 to December 1957.

$$57, 73, 58, 60, 66, 50, 45, 50, 28.$$

Find the median, mode and mean of this group of numbers.
     (L)

7. The mean of one set of six numbers is $8\frac{1}{2}$ and the mean of a second set of eight different numbers is $7\frac{1}{4}$. Calculate the mean of the combined set of fourteen numbers.   (L)

8. Calculate (i) the arithmetic mean, (ii) the median, (iii) the mode of the following numbers.

10, 14, 22, 16, 15, 14, 15, 13, 14, 17.

(*A.E.B.*)

9. A firm employs 20 men, 40 women and 32 boys. A woman's work is considered to be three-quarters as effective as a man's and a boy's work half as effective as a man's. If a man's rate of pay is £12 a day, a woman's and a boy's is proportionate to their effectiveness, calculate the arithmetic mean daily rate of pay for the 92 workers.

10. Define the arithmetic mean, the median and the mode of a distribution, and give an example in which they coincide.

In a certain factory 10 employees earn £60 per week, 35 earn £70 per week, 25 earn £80 per week and 30 earn £90 per week. Find the mean, the median and the mode of this wage distribution.

Which of these will be altered if 4 employees work overtime and each increases his weekly wage by £12?

11. The arithmetic mean of the numbers x, y, and z is 6 and the arithmetic mean of the numbers x, y, z, a, b, c and d is 9. What is the arithmetic mean of the numbers a, b, c and d? (*L*)

12. The following figures give the quarterly totals (in million pounds) of new work done on housing for public authorities for the period April 1959 to June 1961

62   62   62   59   60   59   62   59   65

Find the arithmetic mean, the mode and the median of this group of numbers. (*L*)

13. Find the median, the mode, the arithmetic mean and the geometric mean of the numbers

2, 3, 3, 4, 5, 8, 10.

(*J.M.B.*)

14. Define the geometric mean of two different positive numbers a and b and prove that it is less than their arithmetic mean. (*J.M.B.*)

12 Averages of More Complicated Distributions

The Arithmetic Mean

Example

A form was set a test and the marks awarded were as follows. Find the arithmetic mean mark.

Mark   4   5   6   7   8   9   10

Number awarded mark   2   4   10   15   5   2   2

The total number of marks awarded is

$(2 \times 4) + (4 \times 5) + (10 \times 6) + (15 \times 7) + (5 \times 8) + (2 \times 9) + (2 \times 10)$
$= 271,$

and the total number of pupils is $2 + 4 + 10 + 15 + 5 + 2 + 2$
$= 40.$

Therefore the A.M. is $\dfrac{271}{40} = 6 \cdot 8$.

These figures are not very large, and do not involve long calculations. If the figures are more complicated it is easier to use the simplification *A.M. = W.M. + arithmetic mean of the deviations*, and to arrange the work as follows. Let the working mean be 7.

Col. 1 Value of mark	Col. 2 Frequency	Col. 3 Deviation from Working Mean	Col. 4 Total deviation (Col. 2 × Col. 3)
4	2	-3	-6
5	4	-2	-8
6	10	-1	-10
			-24
7	15	0	
8	5	1	5
9	2	2	4
10	2	3	6
	Total 40		15
			Total -9

Therefore A.M. = W.M. + arithmetic mean of the deviations
$$= 7 - \frac{9}{40} = 7 - 0.225 = 6.775 = 6.8$$

Method
1. Arrange in columns. Col. 1 = value of mark.
   Col. 2 = frequency.

2. 0 is placed opposite the working mean in Col. 3, and opposite the other items a number is placed to show how much the item deviates from the working mean. Thus 5 differs by -2 from 7, and 10 differs by 3 from 7.

3. Multiply Col. 2 by Col. 3 and write the result in Col. 4.

4. Find the sum of Col. 2 and the sum of Col. 4. The sum of Col. 4 is best found by adding all the negative numbers, and all the positive numbers, and finding the final total from these.
   Then, the arithmetic mean = working mean + arithmetic mean of the deviations from the working mean.

Exercise 12.1
Find the A.M. of the following:

1. | Marks | 1 | 2 | 3 | 4 | 5 | 6 | 7 | 8 | 9 | 10 |
   |---|---|---|---|---|---|---|---|---|---|---|
   | Frequency | 2 | 2 | 4 | 6 | 10 | 12 | 10 | 9 | 4 | 1 |

2. | Number of peas in a pod | 4 | 5 | 6 | 7 | 8 | 9 |
   |---|---|---|---|---|---|---|
   | Frequency | 40 | 90 | 110 | 90 | 50 | 20 |

3. | Number of persons per house | 1 | 2 | 3 | 4 | 5 | 6 | 7 | 8 |
   |---|---|---|---|---|---|---|---|---|
   | Frequency | 56 | 124 | 161 | 148 | 73 | 25 | 8 | 5 |

4. | Number of boys per form | 28 | 29 | 30 | 31 | 32 | 33 |
   |---|---|---|---|---|---|---|
   | Frequency | 5 | 9 | 9 | 10 | 10 | 7 |

5. | Height of boys in cm | 160 | 161 | 162 | 163 | 164 | 165 |
   |---|---|---|---|---|---|---|
   | Frequency | 5 | 7 | 8 | 10 | 9 | 1 |

Further Simplification
If the deviations have a common factor, we make use of the second simplification (see page 124) by measuring the deviations in units of the common factor.

Example

Find the arithmetic mean of the following frequency distribution.

Value of Item	5	10	15	20	25	30	35	40
Frequency	2	5	7	9	12	10	3	2

Decide on a working mean, say 25, and arrange the work as follows:

Col. 1 Value of Item	Col. 2 Frequency	Col. 3 Deviation from W.M. in units of 5	Col. 4 Col. 2 × Col. 3
5	2	−4	− 8
10	5	−3	−15
15	7	−2	−14
20	9	−1	− 9
			−46
25	12	0	
30	10	1	10
35	3	2	6
40	2	3	6
	Total 50		22
			Total −24

The sum of Col. 4 is in units of 5. Therefore the correct sum is −24 × 5, and

A.M. = W.M. + arithmetic mean of the deviations from the working mean

$$= 25 + \frac{(-24) \times 5}{50} = 25 - 2 \cdot 4 = 22 \cdot 6$$

Exercise 12.2

Find the arithmetic mean of the following:

1.
Value of Item	5	10	15	20	25	30
Frequency	6	7	9	12	10	6

2.
Value of Item	10	20	30	40	50	60
Frequency	9	12	47	62	49	21

3.
Value of Item	11	22	33	44	55	66
Frequency	7	8	6	9	3	7

4.
Value of Item	13	26	39	52	65	78
Frequency	4	10	16	18	12	10

If the items of a frequency distribution are classified in intervals, we make the assumption that every item in an interval has the mid-value of the interval. The error introduced is negligible if the frequencies are large.

Example

The marks out of a total of 100 obtained by 200 candidates in an examination are classified by intervals of 10 marks. The classes are set out below. Calculate the arithmetic mean.

Marks	1–10	11–20	21–30	31–40	41–50	51–60
Frequency	7	11	14	28	56	43

Marks	61–70	71–80	81–90	91–100
Frequency	23	12	4	2

Col. 1 Class Interval	Col. 2 Centre of Class Interval	Col. 3 Frequency	Col. 4 Deviation from W.M. in units of 10	Col. 5 Col. 3 × Col. 4
1–10	5·5	7	−4	−28
11–20	15·5	11	−3	−33
21–30	25·5	14	−2	−28
31–40	35·5	28	−1	−28
				−117
41–50	45·5	56	0	
51–60	55·5	43	1	43
61–70	65·5	23	2	46
71–80	75·5	12	3	36
81–90	85·5	4	4	16
91–100	95·5	2	5	10
		Total 200		151
				Total 34

Therefore, A.M. = W.M. + arithmetic mean of the deviations from W.M.

$$= 45·5 + \frac{34 \times 10}{200} = 47·2.$$

If the class intervals are unequal we proceed in the same way.

Col. 1	Col. 2	Col. 3	Col. 4	Col. 5
Class	Centre of Class	Frequency	Deviation from W.M.	Col. 3 × Col. 4
1–5	3	40	− 17·5	− 700
6–10	8	80	− 12·5	− 1000
				− 1700
11–30	20·5	100	0	
31–50	40·5	60	20	1200
51–80	65·5	40	45	1800
81–100	90·5	20	70	1400
		Total 340		4400
				Total 2700

Therefore, A.M. = W.M. + arithmetic mean of deviations from W.M.

$$= 20·5 + \frac{2700}{340} = 28·4.$$

Exercise 12.3

Find the arithmetic mean of the following frequency distributions:

1. | Class | 1–5 | 6–10 | 11–15 | 16–20 | 21–25 |
    |---|---|---|---|---|---|
    | Frequency | 3 | 5 | 8 | 6 | 3 |

2. | Class | 1–10 | 11–20 | 21–30 | 31–40 | 41–50 |
    |---|---|---|---|---|---|
    | Frequency | 6 | 8 | 10 | 9 | 7 |

3. The following table gives the number of male workers employed on agricultural holdings in England and Wales, and in Scotland, classified in age groups. Compare the mean age of the workers in England and Wales, and Scotland.

Age	15–18	18–21	21–65	65–70
Workers (Thousands)				
England and Wales	43	33	366	22
Scotland	6	6	53	3

4. When finding the arithmetic mean of a frequency distribution, we make the assumption that all items in a given class have the mid-value of that class. If this assumption is not valid, find the possible limits of the arithmetic mean of the following frequency distribution of a continuous variable.

Class	1–5	6–10	11–15	16–20	21–25
Frequency	4	8	15	13	10

5. Find the arithmetic mean of the following:

Class	1–4	5–8	9–12	13–16	17–20	21–24
Frequency	3	6	10	14	12	5

The median is easily calculated for a small discrete distribution, but in case of a large continuous frequency distribution where the classes are given in intervals, it is better to find the median graphically. First a table is drawn up giving the cumulative frequency, that is, the frequency formed by adding each frequency to the sum of the previous ones.

Example
The following is the frequency distribution of the masses of 100 articles to the nearest gramme, and classified into intervals of 10 grammes.

Mass in grammes	100–109	110–119	120–129	130–139	140–149	150–159
Frequency	2	15	37	31	11	4

Find the median average.

Class	Frequency	Cumulative Frequency	Reverse Cumulative Frequency
100–109	2	2	100
110–119	15	$2+15 =$ 17	98
120–129	37	$17+37 =$ 54	83
130–139	31	$54+31 =$ 85	46
140–149	11	$85+11 =$ 96	15
150–159	4	$96+ 4 = 100$	4

If the cumulative frequency is plotted against the mass, we have a curve which, because of its shape, is known as the ogive. The cumulative frequency graph of a one-humped distribution is always of this shape. Other distributions give cumulative frequency graphs of different shapes. Thus, the cumulative frequency graph of a rectangular distribution is a straight line. Care must be taken to plot the curve correctly. Thus, the cumulative frequency for the third group, 120–129, is 54, and since the masses of the articles are measured to the nearest gramme, the extreme limits of this group are 119·5 and 129·5 g, and 54 is the cumulative frequency to 129·5. Therefore, the cumulative frequencies must be plotted against 109·5, 119·5, 129·5, and so on to 159·5. On doing this, we obtain the following continuous curve.

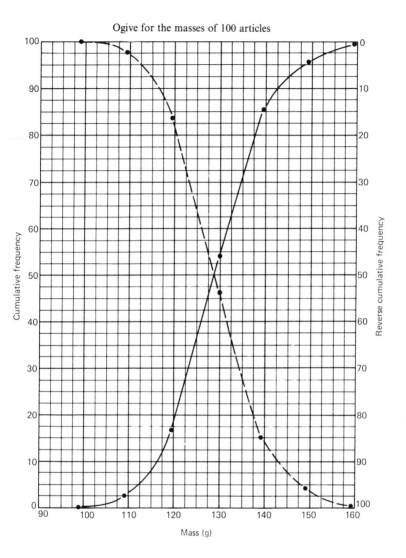

Ogive for the masses of 100 articles

In the case of a continuous distribution, we arrive at the median value of the variable, whether the items are cumulated in ascending or descending order of magnitude, by taking the value of the variable corresponding to the frequency half-way along the cumulative frequency scale. The reason for this becomes evident if we also cumulate the frequencies in the reverse order of magnitude of the variable, and plot these cumulative frequencies on the same graph.

The cumulative frequency down to 149·5 is 4, down to 139·5 is 15, and so on. This curve is represented by the broken line on the diagram. We see that the curves intersect half-way along the cumulative frequency scale and that the median, which is the value of the mass at this point, is 128·5 g.

If we assume the portion of the curve joining two adjacent points is nearly a straight line we can make use of the principle of similar triangles to calculate the value of the median. In the last example the cumulative frequency to the value of the variable at 119·5 is 17, and to the value of the variable at 129·5 is 54. We want the value of the variable where the cumulative frequency is $\dfrac{n+1}{2} = \dfrac{101}{2} = 50\cdot5$.

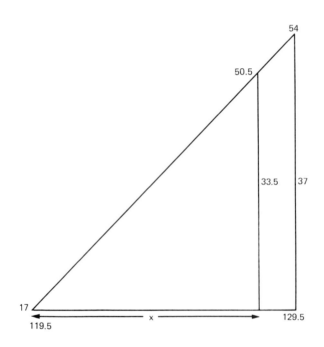

From the similar triangles we have

$$\frac{x}{10} = \frac{33\cdot5}{37}$$

or
$$x = 9\cdot05$$

Therefore, the value of the variable where the cumulative frequency is 50·5 is
$$119\cdot5+9\cdot1 = 128\cdot6$$

Quartiles, Deciles and Percentiles

The median is an average that depends upon its position in the distribution. Other values that depend on their position in the distribution are quartiles, deciles and percentiles. Collectively, these values are known as quantiles or grades.

The lower, or first quartile (Q_1), is one-quarter the way along the distribution, and the upper, or third quartile (Q_3), is three-quarters the way along the distribution. If the cumulative frequency range is divided into ten equal parts, the corresponding values of the variable are known as deciles, and if into one hundred parts, as percentiles. The median corresponds to the second quartile, the fifth decile, or the fiftieth percentile.

As in the case of the median, if the order of the variables is reversed, the lower quartile in the arrangement by ascending magnitude must become the upper quartile in the arrangement by descending magnitude. Thus, in a discrete distribution, Q_1 is the magnitude of the variable corresponding to the $\frac{n+1}{4}$th item, and Q_3 is the value of the variable corresponding to the $\frac{3(n+1)}{4}$th item. In the following discrete distribution, Q_1 is the value of the variable corresponding to the $\frac{6+1}{4} = 1\frac{3}{4}$ item, and if the items are reversed in order of magnitude then Q_3 is the value of the variable corresponding to the $\frac{3(6+1)}{4} = 5\frac{1}{4}$ item, and in this example it would be taken to be $8\frac{1}{2}$ i.e. $[7+\frac{3}{4} \times 2$ or $9-\frac{1}{4} \times 2]$

Value of the variable	7	9	12	15	18	19
Number of item	1	2	3	4	5	6

Position of Q_1. $1\frac{3}{4}$ items ⟶   ⟍ Position of Q_3 if the items are reversed. $5\frac{1}{4}$ items.

Similarly, in a discrete distribution, the lower and upper deciles are the values of the variable corresponding to the $\frac{n+1}{10}$th and $\frac{9(n+1)}{10}$th items, and for percentiles, the lower and upper values are $\frac{n+1}{100}$th and $\frac{99(n+1)}{100}$th items.

If the distribution is continuous, the lower quartile is the value of the variable on the graph corresponding to one-quarter of the distance along the cumulative frequency scale, and the upper quartile to the value

139

of the variable corresponding to three-quarters of the distance along the cumulative scale. Similarly for deciles and percentiles. In the case of a continuous variable, approximate values can be calculated for the quantiles in a manner similar to that illustrated in the case of the median. Deciles are often written as $D_1, D_2, \ldots, D_9$ and percentiles as $P_1, P_2, \ldots, P_{99}$.

Example
The lengths of 1126 articles are measured to the nearest cm, and the frequencies tabulated as follows:

Length in cm	Frequency	Length in cm	Frequency
60	23	67	121
61	68	68	97
62	91	69	71
63	140	70	30
64	180	71	14
65	143	72	4
66	144		

Construct a cumulative frequency table and draw the ogive.
Use the ogive to determine
   (i) the median length of the articles,
   (ii) the difference between the first and third quartiles,
   (iii) the value of the third decile,
   (iv) the value of the 42nd percentile.

Cumulative Frequency Table

Length in cm	Cumulative Frequency	Length in cm	Cumulative Frequency
60	23	67	910
61	91	68	1007
62	182	69	1078
63	322	70	1108
64	502	71	1122
65	645	72	1126
66	789		

Draw the ogive, plotting the corresponding cumulative frequencies against 60·5, 61·5, and so on. On the right-hand side of the graph draw a new percentage scale from 0 to 100.

The median is the 563·5th item, using the left-hand scale, or the value corresponding to the 50th point on the right-hand scale. The value is 64·9 cm.

The lower quartile is the value corresponding to the 281·75th item on the left-hand scale, or the point 25 on the right-hand scale. The value

140

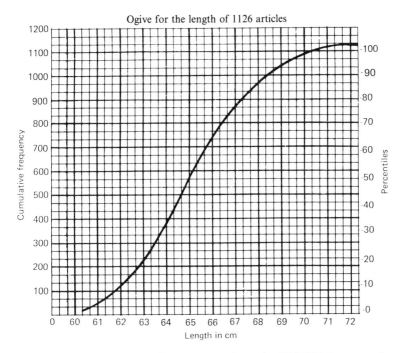

Ogive for the length of 1126 articles

is 63·2 cm. The upper quartile corresponds to the 845·25th item, or the 75 point, on the right-hand scale, i.e. 67 cm. Therefore the difference between the lower and upper quartiles is 3·8 cm.

The third decile is the value corresponding to the 30 mark on the right-hand scale, i.e. 63·6 cm, and the 42nd percentile corresponds to the 42nd mark on the right-hand scale, i.e. 64·3 cm.

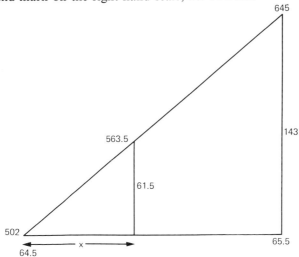

If we wish to calculate the value of the median, we have

$$\frac{x}{1} = \frac{61 \cdot 5}{143} = 0 \cdot 43$$

Median $= 64 \cdot 5 + 0 \cdot 43 = 64 \cdot 9$

The other averages can be calculated in a similar manner.

Exercise 12.4
Draw the ogive and find the median values of the continuous distributions 1 to 5 of Exercise 12.3, page 135.

The Mode
The mode was defined as the most frequent, or most 'popular', value of the variable. Thus in the following example with a discrete variable, the mode 5 is easily distinguished, but in the case of a frequency distribution with class-intervals extended over a range it is more difficult.

Number of Rooms in dwelling	Number of Private House-holds (Thousands)	Number of Rooms in dwelling	Number of Private House-holds (Thousands)
1	161·4	9	211·4
2	795·1	10	109·9
3	1745·6	11	65·0
4	3761·1	12	49·5
5	4567·1	13	25·1
6	1868·9	14	16·9
7	649·4	15 or more	55·2
8	399·9		

In the chapter on frequency distributions is a list of marks (p. 70) which have also been grouped into class-intervals of five. On examining the list of marks, we see that there is no one single mode. The marks 37, 38, 43, 50, 53, each have a frequency of 10. In the grouped list of marks, there is a modal group 36–40. If the distribution is symmetrical, the mode can be assumed to be the mid-value of the modal group, which in this case gives the mode as 38. Furthermore, if the marks are again re-grouped into intervals of ten, we have:

Marks	1–10	11–20	21–30	31–40	41–50	51–60
Frequency	11	24	50	76	71	46

Marks	61–70	71–80	81–90	91–100
Frequency	35	24	12	4

This gives the modal group as 31–40, and, on the above assumption, the mode is 35·5. Thus we see that the value of the mode is entirely dependent on the choice of class-interval. If the class-intervals are large or unequal in size, it is impossible to approximate to the value of the mode, with any degree of mathematical precision.

In an asymmetrical distribution the following method gives a better approximation. In the classification of marks in intervals of five, the class below the modal class has four less candidates than the modal class, and the class above the modal class has three less candidates than the modal class. Therefore assume the mode will divide the modal class in the ratio of 4 to 3; i.e. the mode is probably

$$35·5 + \frac{4}{7} \text{ of } 5 = 38·4$$

When the marks are re-grouped into intervals of ten, the mode is, on the above assumption

$$30·5 + \frac{26}{31} \text{ of } 10 = 38·9$$

In the diagrams on the next page we show how the properties of similar triangles may be used to determine the mode geometrically from the histogram. This is the best way to find the mode of a frequency distribution classified in intervals.

In a slightly asymmetrical uni-modal distribution, the following approximate relation is given as existing between the mode, mean, and median:

$$\text{mode} = \text{mean} - 3(\text{mean} - \text{median})$$

If the mean and the median are calculated for the distribution of marks, they are found to be 44·5 and 42·8, respectively. Using the above relation the mode becomes

$$44·5 - 3(44·5 - 42·8) = 39·4$$

Thus the mode of a frequency distribution is often impossible to calculate with any accuracy, and its value depends on what assumption is made. For general purposes, it is best to assume that the mode divides the modal group in the ratio of the differences between the modal group and the groups below and above the modal group.

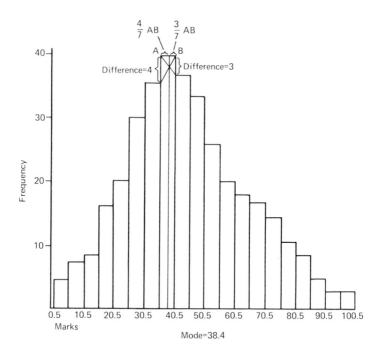

Mode=38.4

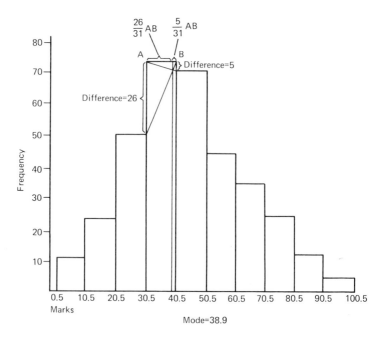

Mode=38.9

144

Note: Since the marks are classed 1–5, 6–10, and so on, the class-intervals on the horizontal scale must go from 0·5 to 5·5, 5·5 to 10·5, 10·5 to 15·5, and so on, so that 3, 8, etc. are the mid-values of the class-intervals.

Example

The following table gives the approximate numbers of males and females who married in a certain year, classified by age. Find the mode of each distribution. (The figures are in thousands.)

Age (years)	Male	Female	Age (years)	Male	Female
15 under 20	11	53	40 under 45	10	9
20 under 25	107	140	45 under 50	6	5
25 under 30	138	98	50 under 55	5	3
30 under 35	51	32	55 under 60	5	3
35 under 40	18	12	60 and over	6	2

As we are only interested in the modal group and the group preceding and following it, we need only draw these three groups in the histogram.

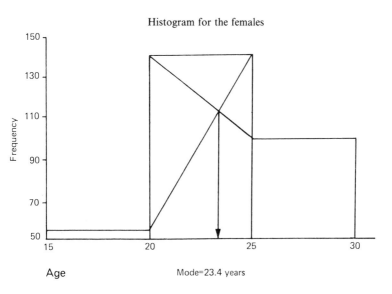

Histogram for the females

Mode=23.4 years

145

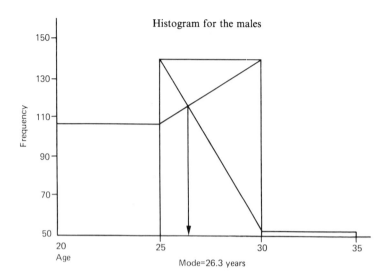

Histogram for the males

Mode=26.3 years

Exercise 12.5

Calculate the mode for the frequency distributions for the examples 1 to 5 of Exercise 12·3 on page 135.

Desirable Properties of an Average

From our study of averages we conclude that it is desirable that an average should be:
1. defined so that each observer arrives at the same result; that is, nothing should be left to the estimation of the observer.
2. based on all the items of the group.
3. quickly and easily understood; that is, it should possess simple and obvious properties.
4. such that the average of a sample of the distribution is reasonably equal to the average of the whole.
5. such that it lends itself to algebraical treatment.

Comparison of the Averages

The Arithmetic Mean
Advantages:
1. is simply calculated.
2. takes into consideration the value of every item.

3. properties are easily understood.
4. in most cases least affected by errors of sampling.
5. particularly easy to manipulate algebraically.

Disadvantages:
1. badly affected by abnormal individual items.

The Median

Advantages:
1. more easily calculated than the mean.
2. unaffected by extreme abnormal values.
3. its properties are easily understood.
4. can be obtained even when the values of extreme items are unknown.
5. can be used for measuring qualities and factors to which mathematical measurements cannot be given.

Disadvantages:
1. does not take into consideration the value of every item.
2. is difficult and almost impossible to use algebraically.
3. if the items are few it is not likely to be representative.
4. the arranging of data in the necessary array can be tedious.

The Mode

Advantages:
1. represents an important value of the variable.
2. is unaffected by abnormal individual items.
3. its properties are easily understood.

Disadvantages:
1. does not take into consideration the values of all the items.
2. is difficult and sometimes impossible to determine with accuracy.
3. is unsuited for algebraical work.

The Geometric Mean

Advantages:
1. takes into account the values of all the items.
2. is determined with accuracy.
3. is fairly easily calculated.
4. is in certain cases easily treated algebraically.
5. reduces the effect of a single large item.

Disadvantages:
1. has no simple properties to make it easily understood.
2. impossible to use when any items are zero.

It is generally accepted that the arithmetic mean is the best form of average for all general purposes, and should always be employed unless there is very definite reason for the choice of another form of average.

Exercise 12.6

Obtain the mean, median, and mode of the following distribution and test the accuracy of the statement,

$$\text{Mode} = \text{Mean} - 3(\text{Mean} - \text{Median}).$$

1.
Daily earnings	£5	£6	£7	£8	£9	£10	£11	£12	£13	£14
No. of persons	20	72	110	173	246	357	296	74	13	9

2.
No. in family	1	2	3	4	5	6	7
No. of families	689	2240	2463	1987	1268	784	455

No. in family	8	9	10 or more
No. of families	215	193	97

3.
Runs scored	1–5	6–10	11–15	16–20	21–25	26–30
Frequency	23	7	8	9	5	4

Runs scored	31–35	36–40	41–45
Frequency	3	3	1

4. From your newspaper find the frequency distributions of goals scored by soccer teams for two wet weeks and two dry weeks. Compare the mean, mode, and median in each case.

Exercise 12.7

1. (a) The register of students taking a college course gave the age groups set out below. Calculate the mean age of the students.

Age	19–20	20–21	21–22	22–23	23–24
No. in each group	2	7	12	6	4

(b) Explain what is meant by the 'mode' of a frequency distribution. State the conditions under which the mode can be found (i) exactly, (ii) with difficulty, (iii) not at all. (L)

2. The following table shows the frequency distribution of the marks of 800 candidates at an examination. Construct a table showing the cumulative frequency distribution and draw a graph of the ogive. Find the mode, the arithmetic mean and the median of the distribution.

Marks	1–10	11–20	21–30	31–40	41–50
No. of candidates	10	40	80	140	170

Marks	51–60	61–70	71–80	81–90	91–100
No. of candidates	130	100	70	40	20

(L)

3. The following table shows the weekly wages earned by the 85 employees in a certain factory. Using a suitable working mean, or otherwise, calculate the average weekly wage as an arithmetic mean.

Weekly wage (in £)	30–39	40–49	50–59	60–69	70–79
No. of wage-earners	3	17	33	24	8

4. Distinguish between median and mode.

In a savings group there are 800 members, and the number of savings certificates held by them is shown in the following frequency table. Construct a table showing the cumulative frequency distribution and draw the graph of the ogive. Find (a) the mode, (b) the median and (c) the semi-interquartile range.

Number of Certificates held	1–50	51–100	101–150	151–200	201–250
Number of members	5	10	30	60	100

Number of Certificates held	251–300	301–350	351–400	401–450	451–500
Number of members	140	220	150	50	35

(L)

5. The following table gives the frequency distribution of marks obtained by 130 candidates in each of two subjects **A** and **B**. Construct a table showing the cumulative frequency distributions in each subject and draw in one diagram the graphs of their ogives. From your diagram, find

(a) the percentage number of candidates that fail in each subject if the pass mark in subject **A** is 55 and that in subject **B** is 35;

(b) the median mark in each subject.

Percentage	1–10	11–20	21–30	31–40	41–50	51–60
Subject **A**	0	0	1	3	6	24
Subject **B**	5	26	30	28	25	9

Percentage	61–70	71–80	81–90	91–100
Subject **A**	30	31	22	13
Subject **B**	5	0	1	1

(L)

6. The following is the frequency distribution of the weights of 100 articles:

149

Wt. in newtons	Frequency	Wt. in newtons	Frequency
100–109	1	160–169	17
110–119	2	170–179	10
120–129	5	180–189	6
130–139	11	190–199	4
140–149	21	200–209	2
150–159	20	210–219	1

Construct a histogram to illustrate these facts. Find the mode of the distribution. (L)

7. Draw a cumulative frequency curve of the data given in Question 6. From your curve find the median weight of the articles, and the percentage of articles in the group weighing over 168 N. (L)

8. The following table gives details of weekly income and of weekly expenditure on food of a group of four families.

Family	Total family income (£)	Expenditure per head on food (£)	Total number in family
A	76	8	4
B	120	9	6
C	85	7·5	5
D	175	11	4

Find to the nearest 5p the average income per head of the group and the average expenditure on food per family.

Which families spend (a) more, (b) less than the average, and by how much?

9. In a botanical experiment, the lengths of 100 leaves from a certain species of plant were measured and the frequencies tabulated as follows:

Length of leaf in mm	Frequency	Length	Frequency
20–24	1	45–49	21
25–29	5	50–54	15
30–34	10	55–59	3
35–39	19	60–64	1
40–44	25		

Construct a histogram to show the frequency distribution and state the mode. (L)

10. From the data of Question 9, construct a cumulative frequency table and draw the ogive. Use the ogive to determine (i) the median length of leaf and (ii) the difference in length between the first and third quartiles. (L)

11. Soil samples taken over an area of heathland classified according to acidity (pH):

Acidity (pH)	4·8–	5·0–	5·2–	5·4–	5·6–	5·8–	6·0–	6·2–
Percentage of sample	7·4	11·2	19·3	25·7	20·1	12·1	4·2	

Construct a cumulative percentage frequency graph for the distribution of soil samples according to acidity given in the table above, and estimate the median and 25th percentile. (*J.M.B.*)

12. The following table shows the frequency distribution of the examination marks of 700 candidates in a Mathematics examination.

Examination marks	10–19	20–29	30–39	40–49	50–59
No. of candidates	10	32	45	87	117
Examination marks	60–69	70–79	80–89	90–99	
No. of candidates	148	127	90	44	

Taking a working zero in the class 60–69 and a unit of 10 marks, calculate the arithmetic mean.

Construct a cumulative frequency curve and use it to determine:
(*a*) the percentage of candidates passing if the pass mark is 55,
(*b*) the lowest mark for distinction if 5 per cent of the candidates are to be given distinction. (*J.M.B.*)

13. The following table shows the marks obtained by the same 150 candidates in two different papers in a recent examination.

	0–10	11–20	21–30	31–40	41–50
No. of candidates (Paper 1)	2	6	20	28	31
No. of candidates (Paper 2)	3	12	22	31	33
	51–60	61–70	71–80	81–90	91–100
No. of candidates (Paper 1)	20	17	11	8	7
No. of candidates (Paper 2)	15	14	9	9	2

Estimate the mean and median marks for Paper 1 only. If you were told that the true median mark for Paper 1 was in fact 48, what conclusion could you draw about the grouping of the marks? Comment on the relative difficulty of the two papers. (*C.A.*)

14. The following table shows the mean quarterly expenditure on capital goods for households in three districts, the households being classified according to whether or not they read the local paper.

	District 1		District 2	
	Readers	Non-readers	Readers	Non-readers
No. of households	8	4	19	8
Mean expenditure (£)	11	13	14	16

	District 3	
	Readers	Non-readers
No. of households	14	7
Mean expenditure (£)	12	11

Calculate the total and mean expenditures for all the households reading the local paper and for all not doing so.

Comment on the following extract from the local paper:

'£50 a year in every household available to advertisers. A recent survey shows that over two-thirds of the total local expenditure on capital goods is by readers of this paper. In the next three months every household taking this paper will spend over £12.70 on capital goods. This money will be spent in *your* shop if you advertise in this paper.' (C.A.)

15. The heights of 574 boys are measured correct to the nearest centimetre, and the frequencies tabulated as follows:

Height (cm)	160	161	162	163	164	165	166	167	168	169	170
Frequency	12	34	45	70	90	83	71	60	48	35	15

Height (cm)	171	172
Frequency	7	4

(i) Construct a histogram to illustrate these facts, and find the mode of the distribution.

(ii) Draw a cumulative frequency curve of the data and from your curve find the median height of the boys. (A.E.B.)

16. A machine produces rods whose diameters are required to be within the tolerance limits 9·88 mm to 10·12 mm. A sample of 150 rods, measured to the nearest hundredth of a millimetre, gave the following distribution:

Diameter (mm)	9·76 to 9·81	9·82 to 9·87	9·88 to 9·93	9·94 to 9·99	10·00 to 10·05
Frequency	1	5	30	71	34

Diameter (mm)	10·06 to 10·11	10·12 to 10·17
Frequency	7	2

(i) Construct the cumulative frequency curve for the rods.

(ii) Calculate the percentage number of rods outside the tolerance limits. (*A.E.B.*)

17. The table shows the distribution of Annual Incomes (after tax) between £750 and £10,000 in U.K. for 1974/5.

Draw the cumulative frequency curve for incomes under £4000 and find the median and lower and upper quartiles for the complete range of incomes given.

Income (£)	No. of Incomes (Thousands)
750 under 1000	2057
1000 under 1500	4466
1500 under 2000	4208
2000 under 3000	6343
3000 under 4000	2502
4000 under 6000	903
6000 under 8000	174
8000 under 10000	52

18. The following frequency distribution was drawn up from records kept over a long period for a machine which was supposed to produce components of dimension 20 mm;

Dimension of component (mm)	Percentage
less than 19·85	1
19·85–19·90	2
19·90–19·95	10
19·95–20·00	25
20·00–20·05	32
20·05–20·10	21
20·10–20·15	8
20·15 and over	1

(i) Calculate the mean dimension of components produced by the machine.

(ii) Draw a cumulative frequency graph and use it to estimate the 10–90 percentile range of dimensions.

In order to check that the machine was maintaining its standard of production 20 components from it were measured accurately and their dimensions were as follows

20·04 20·10 20·06 19·98 19·98 20·04 20·12 20·00 20·08 20·06
19·92 20·04 19·98 20·02 20·00 19·96 20·06 19·88 19·98 20·10

Calculate the mean of this sample and state how many of the components of the sample had dimensions outside the established 10–90 percentile range. Comment on any conclusion that can be drawn from the results. (*J.M.B.*)

13 Moving Averages

Moving Average

If we have a series of observations, and the first and second term are averaged, then the second and third term, then the third and fourth term, and so on, we have a two point moving average. If the terms are averaged in this way, three at a time, we have a three point moving average, and so on. *The calculation of moving averages is a means of smoothing out irregularities in a graph so that variations and trends over a period of time become more evident.*

Example

Table of Employment Vacancies Unfilled in Great Britain for Four Years
(Thousands)
End of period

Year	Jan.	Feb.	Mar.	Apr.	May	June
1	222	233	263	279	281	296
2	261	258	288	326	331	381
3	338	352	371	416	426	443
4	373	368	380	390	380	

Year	July	Aug.	Sept.	Oct.	Nov.	Dec.	
1	321	295	282	276	272	270	
2	383	356	348	340	333	338	
3	473	448	424	401	388	382	(*A.*)

When a graph is plotted to illustrate the figures in the above table we obtain an irregular outline, but it does suggest there may be a seasonal variation. The seasonal variation can be better illustrated if the irregularities in the graph are smoothed out. This is done as follows: average the figures for the first six months, January to June, then for the six months February to July, then the figures for the six months March to August, and so on; i.e. calculate a six point moving average. Plot each of the averages at the middle point of the period it represents. Thus the first average is plotted half-way between the end of March and the end of April, the second average half-way between the end of April and the end of May, and so on. This gives us a fairly smooth curve, and the seasonal variation is very noticeable.

If the process is repeated, taking a twelve-month moving average, the seasonal variation is eliminated and the general trend or, as it is sometimes known, the secular trend, becomes more noticeable. There is an increase in the number of unfilled vacancies over the period.

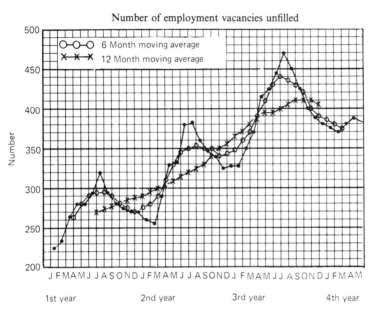

Number of employment vacancies unfilled

Simplification

It can be a laborious process calculating a moving average if the values of the items are large. Fortunately, the process can be simplified. For if the values of the items are, a_1, a_2, and so on, and it is necessary to calculate a three point moving average, we have

$$\text{1st average} = \frac{a_1 + a_2 + a_3}{3} = m_1$$

$$\text{2nd average} = \frac{a_2 + a_3 + a_4}{3} = \frac{a_1 + a_2 + a_3 + a_4 - a_1}{3} = m_1 + \frac{a_4 - a_1}{3} = m_2$$

$$\text{3rd average} = \frac{a_3 + a_4 + a_5}{3} = \frac{a_2 + a_3 + a_4 + a_5 - a_2}{3} = m_2 + \frac{a_5 - a_2}{3} = m_3$$

i.e. to find the next average, we add one-third of the difference between the next term and the first term of the previous average. If we were finding a five point moving average we would add one-fifth of the difference between the next term and the first term of the previous average, and so on.

Example

To calculate the five-point moving average, it is best to arrange the work as follows:

Items	Working	Average
4		
7		
3	$\dfrac{25}{5} = \quad 5\cdot0$	5·0
9	$\dfrac{1-4}{5} = -0\cdot6$	4·4
2	$\dfrac{6-7}{5} = -0\cdot2$	4·2
1	$\dfrac{4-3}{5} = \quad 0\cdot2$	4·4
6	$\dfrac{5-9}{5} = -0\cdot8$	3·6
4	$\dfrac{7-2}{5} = \quad 1\cdot0$	4·6
5	$\dfrac{9-1}{5} = \quad 1\cdot6$	6·2
7	$\dfrac{8-6}{5} = \quad 0\cdot4$	6·6
9	$\dfrac{10-4}{5} = \quad 1\cdot2$	7·8
8		
10		

Method

In column 1 place the items.

In column 2 do the working. Average the first five items, and place the value opposite the middle item of the average in column 3.

For the second average, subtract the first item from the sixth item, divide by 5 and add the result to the previous average. The working is shown in column 2. For the next average subtract the second from the seventh item, divide by 5 and add the result to the previous average and so on.

Example

The mean hours of sunshine per day for two years are given in the following table. Plot the given figures, and on the same graph plot the six-month and twelve-month moving averages.

Sunshine

Mean Hours Per Day

Year	Jan.	Feb.	Mar.	Apr.	May	June
1	1·68	2·37	4·06	6·24	5·58	6·01
2	1·60	2·59	3·43	6·13	6·28	5·79

Year	July	Aug.	Sept.	Oct.	Nov.	Dec.	
1	5·64	5·17	3·11	3·18	1·45	1·26	
2	6·11	4·66	3·55	2·67	1·28	1·89	(A.)

Some of the working is shown to illustrate the method.

Mean Hours	Working for 6 Month Moving Average	6 Month Average	Working for 12 Month Moving Average	12 Month Average
1·68				
2·37				
4·06				
	$\dfrac{25\cdot94}{6} =$ 4·32	4·32		
6·24				
	$\dfrac{5\cdot64-1\cdot68}{6} =$ 0·66	4·98		
5·58				
	$\dfrac{5\cdot17-2\cdot37}{6} =$ 0·47	5·45		
6·01				
	$\dfrac{3\cdot11-4\cdot06}{6} = -0\cdot16$	5·29	$\dfrac{45\cdot75}{12} =$ 3·813	3·81
5·64				
	$\dfrac{3\cdot18-6\cdot24}{6} = -0\cdot51$	4·78	$\dfrac{1\cdot60-1\cdot68}{12} = -0\cdot007$	3·81
5·17				
3·11				
3·18	and so on.			
1·45				
1·26				
1·60				

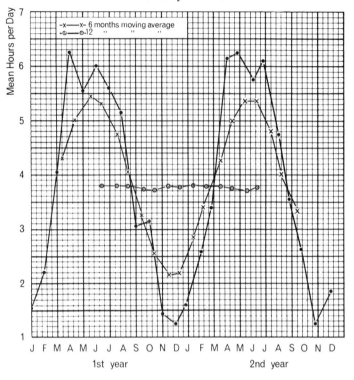

Mean daily sunshine

Variations

The purpose of calculating moving averages is to eliminate unwanted variations. The common variations are:

1. Secular variation or general trend.
   This is the long-term trend, and in our first example it was for the number of unfilled vacancies to increase; while in the second example, for the annual amount of sunshine to remain constant.

2. Seasonal variation.
   This is due to seasonal causes, and in the first example it shows there are more unfilled vacancies in the summer than in the winter.

3. Cyclical variation.
   This is a periodic variation and may take a number of years to work itself out, as in the eleven-year period for maximum sunspot activity.

4. Special or random variations.

These are due to accidental or occasional happenings, such as a strike or a political crisis which may temporarily affect many aspects of the life of the nation.

The span of a moving average will depend on the variation to be removed; the greater the span, the smoother the graph. The general trend is generally best shown by a twelve-month moving average, and a seasonal variation by a three or six-month moving average. To eliminate a variation, the span of the moving average should be a multiple of the number of observations that make a period.

Seasonal Adjustment

When comparing the values of a variable in two successive quarters, it is useful to know the regular seasonal fluctuations and to be able to make allowance for them. Seasonal fluctuations are not perfectly regular and therefore not predictable with certainty. We can calculate an average seasonal fluctuation, and this is what we mean when we talk of a typical seasonal fluctuation. We regard the trend value of a time series as the normal value, and the actual value as a deviation from the normal caused by the regular seasonal deviation together with any other irregular factors. Consider the following example from an A.E.B. paper.

Example

The following table gives the takings (in £ Thousands) of a shopkeeper in each quarter of four successive years. Draw a graph to illustrate the data and on it superimpose a graph of the four-quarterly moving average.

State the conclusions to be drawn from the graphs.

Year	1st qr.	2nd qr.	3rd qr.	4th qr.
1	13	22	58	23
2	16	28	61	25
3	17	29	61	26
4	18	30	65	29

Calculating the four-quarterly moving averages, we have:
29 29·75 31·25 32 32·5 32·75 33 33 33·25 33·5 33·75 34·75 35·5
Plotting the original values and the moving averages on the same graph, and remembering that the first average is from the end of the 1st quarter to the end of the 4th quarter and is therefore plotted half-way between

these points, i.e. half-way between the end of the 2nd and 3rd quarters, and so on, we obtain the graphs below.

The conclusions to be drawn are that there is a seasonal fluctuation and that the general trend is slightly upwards.

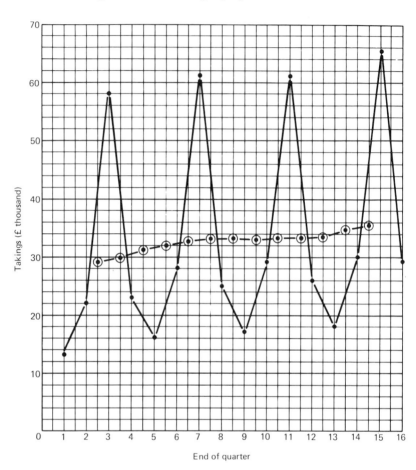

End of quarter

Let us consider the problem further.

By interpolation from the graph, the values of the general trend at the 3rd, 7th and 11th quarters are 29·4, 32·9 and 33·4. The actual values of the variable are 58, 61 and 61, and so the deviations from the general trend, or normal value, are $58 - 29·4 = 28·6$, $61 - 32·9 = 28·1$ and $61 - 33·4 = 27·6$. These are known as the absolute deviations. In practice, it is usual to calculate the relative deviations, that is the ratio of the actual value of the variable to the trend value, and if this is

160

expressed as a percentage it is known as the percentage deviation. The relative deviations for the 3rd, 7th and 11th quarters in this example are $\dfrac{58}{29\cdot4} = 1\cdot97$; $\dfrac{61}{32\cdot9} = 1\cdot85$; and $\dfrac{61}{33\cdot4} = 1\cdot83$, and the average relative deviation is $\dfrac{1\cdot97 + 1\cdot85 + 1\cdot83}{3} = 1\cdot9$. Therefore, to obtain the value of the general trend for the 15th quarter we must divide the value of the takings for this quarter by $1\cdot9$. This gives us $\dfrac{65}{1\cdot9} = 34\cdot2$. Since, in a time series, the trend value is regarded as the normal value, $34\cdot2$ is the 'seasonally adjusted' figure for the 15th quarter. A similar adjustment can be applied to other quarterly figures beyond the general trend values.

Since the moving averages do not coincide with the original values, we have to interpolate from the graph to obtain the seasonal fluctuations. If we choose a three-quarterly moving average, (which in this case is wrong), the first moving average would be plotted at the end of the 2nd quarter and the moving averages would then coincide with the dates of the original series. The deviations could then be found by simple subtraction. The non-coincidence with a date of the original series always occurs if the span of the moving average consists of an even number of items, but the difficulty does not arise if the span consists of an odd number of items. To overcome the difficulty of non-coincidence with a moving average consisting of an even number of items, the moving averages are 'centred'. Thus, if we average the 1st and 2nd values of the moving averages, we can plot this point at the end of the 3rd quarter, and the average of the 2nd and 3rd moving averages can be plotted at the end of the 4th quarter, and so on. In this case there is no need to draw a graph.

To calculate the seasonally adjusted figures, we can now proceed to compile the table on p. 162. From this, it can be seen that the average relative deviation for the 3rd, 7th and 11th quarters is $\dfrac{1\cdot97 + 1\cdot87 + 1\cdot83}{3} = 1\cdot89$; for the 4th, 8th and 12th quarters, it is $\dfrac{0\cdot75 + 0\cdot76 + 0\cdot77}{3} = 0\cdot76$; for the 5th, 9th and 13th quarters, it is $0\cdot52$; and for the 6th, 10th and 14th quarters, it is $0\cdot87$. To obtain the seasonally adjusted figure for the 15th quarter, we must divide the value of the takings for this quarter by $1\cdot89$, i.e. $\dfrac{65}{1\cdot89} = 34\cdot4$; for the 16th quarter we get $\dfrac{29}{0\cdot76} = 38\cdot2$, and so on.

Quarter	Value of Takings (£ Thousand)	4-quarterly moving average	Centred Value	Relative Deviation
1	13			
2	22			
		29		
3	58		29·375	$\dfrac{58}{29·375} = 1·97$
		29·75		
4	23		30·5	$\dfrac{23}{30·5} = 0·75$
		31·25		
5	16		31·625	0·51
		32		
6	28		32·25	0·87
		32·5		
7	61		32·625	1·87
		32·75		
8	25		32·875	0·76
		33		
9	17		33	0·52
		33		
10	29		33·125	0·88
		33·25		
11	61		33·375	1·83
		33·5		
12	26		33·625	0·77
		33·75		
13	18		34·25	0·53
		34·75		
14	30		35·125	0·85
		35·5		
15	65			
16	29			

	Quarter			
	1	2	3	4
Year 1	—	—	1·97	0·75
Year 2	0·51	0·87	1·87	0·76
Year 3	0·52	0·88	1·83	0·77
Year 4	0·53	0·85	—	—
Total	1·56	2·60	5·67	2·28
Average	0·52	0·87	1·89	0·76

Warning

If any fluctuation differs by a large amount from that which might be expected, it may be caused by some exceptional irregular variation and should be left out of the calculation. Thus, in our example, a prolonged local strike could affect the purchasing power of the shopkeeper's customers and alter some of the fluctuations. Not all seasonal adjustments are reliable and care must be taken in using a series of statistics whose seasonal pattern is liable to alteration. These seasonal irregularities are not eliminated by the above corrections.

Exercise 13.1

1. Calculate the five point moving average for the following sets of numbers:
   (*a*) 4 6 8 3 6 8 8 9 9 10 12.
   (*b*) 21 22 18 17 16 10 8 9 8.
   (*c*) 19·2 20·1 21·2 23·6 24·2 25·3 25·7 26·2 26·9 27·3 27·5 27·8 28·4 28·7 29·1 29·8.

2. Explain what is meant by (*a*) secular variation (*b*) seasonal variation in a time series.

   The following table gives the monthly values of the interim index of industrial production from July 1974 to June 1975. Smooth the series by calculating to one decimal place, the three-monthly moving averages and draw a graph of the new series. State any deduction that can be made.

1974	July	Aug.	Sept.	Oct.	Nov.	Dec.
	100	97	109	111	113	103
1975	Jan.	Feb.	Mar.	Apr.	May	June
	103	108	107	104	102	101

3. Explain what is meant by 'moving average'

   The following table gives the numbers (in thousands) of passenger cars and chassis produced for export in the months from April, 1948 to June, 1950. Construct a table showing the twelve-monthly moving averages (correct to the nearest thousand) and draw a graph to show the trend.

	Jan.	Feb.	Mar.	Apr.	May	June
1948				20	20	25
1949	18	19	26	18	21	26
1950	32	33	41	29	34	40

	July	Aug.	Sept.	Oct.	Nov.	Dec.
1948	18	16	25	18	17	19
1949	16	20	26	24	28	28	(*L*)

4. The monthly value (in million of pounds) of exports from the United Kingdom to North America from July 1974 to June 1975 were as follows:

July	Aug.	Sept.	Oct.	Nov.	Dec.
191	200	201	192	190	202

Jan.	Feb.	Mar.	Apr.	May	June
222	194	202	182	195	175

SOURCE: Dept. of Trade

Draw a graph to illustrate these figures.

Calculate the four-monthly centred moving averages to one place of decimals and plot these on the graph.

5. The following table shows the number of new factory buildings completed in twelve consecutive quarters:

	First Year	Second Year	Third Year
First Quarter	256	241	282
Second Quarter	284	248	272
Third Quarter	218	271	289
Fourth Quarter	264	279	294

Draw a graph of the data and of the twelve-monthly moving average. (L)

6. The following table gives the monthly totals (in hundreds) of hire-purchase contracts for new motor cycles during 1957:

Jan.	Feb.	Mar.	Apr.	May	June
78	78	108	124	123	83

July	Aug.	Sept.	Oct.	Nov.	Dec.
81	79	49	52	42	25

(a) Draw a graph to illustrate this data.

(b) Smooth the series by calculating the three-monthly moving averages and superimpose the graph of these moving averages on the first graph.

(c) Find the greatest and the least deviation from the mean. (L)

7. Total personal expenditure of the United Kingdom on consumer goods and services (£ Million):

	1972	1973	1974	1975
1st Quarter	8871	10403	11237	13580
2nd Quarter	9710	10888	12429	
3rd Quarter	10058	11379	13177	
4th Quarter	10833	12138	14439	

Plot a graph to illustrate the above time series. Sketch in the diagram also the straight line which indicates the four quarter centered moving average general trend of the series.

8. A well-known ballet company gave a six-weeks' season at a large hall capable of seating 4000 people and the attendances in hundreds, correct to the nearest hundred, are recorded in the following table:

			Week			
	1	2	3	4	5	6
Monday	12	22	30	40	38	32
Tuesday	13	25	31	40	39	32
Wednesday	20	31	40	40	40	34
Thursday	19	31	38	40	40	32
Friday	17	26	36	40	40	30
Saturday	24	34	40	40	40	36

Plot a graph to illustrate the above time series and sketch in the straight line which indicates the trend during the first three weeks.

Comment on the weekly *cycle* of attendances and state, with reasons, whether or not you think an extension of the season to eight weeks would have been justified. (*J.M.B.*)

9. Explain the purpose of a moving average graph.

The table gives the brightness of a star measured every two days over a period of 64 days.

Days 0–18	11	16	21	24	26	25	18	7	0	3
Days 20–38	8	13	17	21	24	26	22	14	10	11
Days 40–58	13	15	17	20	22	24	23	20	16	12
Days 60–64	12	14	17							

Draw a graph showing the variations in the brightness of the star. Taking a period suitably chosen to smooth out the large periodic variation, and using the same axes, draw a moving average graph. Comment on your graph. (*CA.*)

10. The following table gives the quarterly sales (in thousands) of a certain model of car over a period of four years.

Year	Quarter			
	Jan.–March	April–June	July–Sept.	Oct.–Dec.
1974	25	26	23	18
1975	21	21	18	12
1976	16	16	16	9
1977	13	12	10	6

Plot these values on a graph and on the same axes plot a moving average with period chosen to remove the seasonal variation.

(*CA*.)

11. The following table gives the indices of weekly wage rates and hourly wage rates for sixteen consecutive quarter years.

31 January 1956 = 100

	Indices of Weekly Wage Rates				Indices of Hourly Wage Rates			
	1958	1959	1960	1961	1958	1959	1960	1961
March	113	117	119	124	113	117	121	129
June	113	117	120	125	114	117	122	130
September	115	117	121	125	115	118	124	131
December	116	118	122	126	117	118	126	132

(i) Calculate and tabulate the four-quarterly moving average for both sets of figures.
(ii) Plot both moving averages on the same graph.
(iii) Comment briefly on the most prominent features revealed by the graph. (*A.E.B.*)

12. The table gives the total assets of Discount Houses at the end of each quarter-year.

Assets of Discount Houses
(£ Million)

Period	1960	1961	1962
1st quarter	959	913	972
2nd quarter	990	947	
3rd quarter	954	937	
4th quarter	1054	1077	

(Monthly Digest of Statistics)

Draw up a table giving the four-quarterly moving averages.
Draw a graph to illustrate the data in the table and superimpose upon it a graph of the four-quarterly moving averages.

Assuming the trend continues as indicated from the first quarter of 1960 to the first quarter of 1962, estimate from your graph the seasonal deviation for the first quarter of 1960, 1961 and 1962.

(*A.E.B.*)

13. The table shows the amount of money, in million pounds, spent upon public transport in the United Kingdom during the periods given. Draw a graph to illustrate the data in the table and superimpose on it a graph to show the general trend.

Money spent on Public Transport in United Kingdom.
(Million pounds)

	1970	1971	1972	1973
First quarter	96	113	123	128
Second quarter	112	129	139	143
Third quarter	123	136	150	157
Fourth quarter	117	130	133	152

14. Plot the information given in the table.

Year	1946	1947	1948	1949	1950
Value of variable (End of year)	12·2	11·6	10·6	10·0	11·3

1951	1952	1953	1954	1955	1956	1957	1958	1959
12·4	13·8	12·6	11·6	11·0	11·7	13·2	14·6	13·0

1960	1961	1962	1963	1964	1965	1966	1967
12·5	11·6	12·7	13·8	15·6	14·8	13·8	13·0

(i) From your graph determine the period of cyclical fluctuations.
(ii) Calculate the trend values and plot these values on the same graph.
(iii) Calculate the mean deviation of the cyclical fluctuations.

(*A.E.B.*)

15. The quarterly profits (in £ Thousands) of an airline company are shown for the years 1960–1963.

	1st qr.	2nd qr.	3rd qr.	4th qr.
1960	32	56	60	40
1961	44	88	92	60
1962	60	116	120	80
1963	76	148	156	96

Draw up a table giving the 4-quarterly moving averages. Illustrate your results by a graph showing the quarterly profits and the secular trend. Make a forecast of the total profits for 1964.

(L)

16. The following table gives the annual circulation (in thousands of issues) of a library for the years 1966–77.

Year	1966	1967	1968	1969	1970	1971	1972
Circulation	14·5	15·5	13·8	15·4	16·8	15·6	14·3

	1973	1974	1975	1976	1977
	15·9	17·2	16·8	19·3	18·7

Construct a graph to illustrate these figures and superimpose on it the graph of the four-yearly moving average.

Make use of the latter graph to estimate the circulation for 1978.

(L)

17. The following table gives the Retail Price Index of Fruit and Vegetables for the end of January, April, July and October for seven years. What are the seasonally adjusted figures for January and April of the first year and July and October of the seventh year?

Year	1	2	3	4	5	6	7
January	100	103	104	110	132	128	150
April	113	120	120	130	137	132	135
July	114	124	136	160	162	160	165
October	100	93	104	123	119	141	144

14 Index Numbers

Relatives

If the value of an item at two different times is to be compared, it can be done by either:
1.   quoting the two values of the item, or
2.   quoting the fraction one value is of the other, or
3.   an 'index number', that is, the ratio of the two values expressed as a percentage.

The last method is by far the most convenient.

For example, in 1957, the value of the gross output of agriculture was £1386·2 million and, in 1967, £1864·5 million. Thus, if 1957 is taken as base, the index figure for 1967 is

Example
$$\frac{1864 \cdot 5}{1386 \cdot 2} \times 100 = 134 \cdot 5.$$

The base year is denoted by equating it to 100. In the above example, the base year is denoted by 1957 = 100.

The index figure is known as a 'relative', and in this case as an agricultural output relative. If the price of a commodity at two different times is compared, the index figure is then known as a price relative, and if wages are compared, as a wages relative.

The index figure above compares the gross agricultural output in 1967 with that in 1957. If the gross output in 1957 is to be compared with that in 1967, the 1967 output is taken as base and the index figure is then
$$\frac{1386 \cdot 2}{1864 \cdot 5} \times 100 = 74 \cdot 3.$$

Find, to the nearest whole number, the price relative for a commodity,
 (i) taking 1952 as base,
(ii) taking 1970 as base.

Year	1952	1970
Price of Commodity	£2·76	£3·83

 (i) We must calculate the 1970 price as a percentage of the 1952 price. The work is best arranged as follows:

1952	1970
100	x
£2·76	£3·83

Since these quantities are to be in proportion, we have

$$\frac{x}{100} = \frac{3 \cdot 83}{2 \cdot 76}$$

$$x = \frac{3 \cdot 83 \times 100}{2 \cdot 76} = 138 \cdot 8$$

(ii) We must calculate the 1952 price as a percentage of the 1970 price.

1952	1970
x	100
£2·76	£3·83

$$\frac{x}{100} = \frac{2 \cdot 76}{3 \cdot 83}$$

$$x = \frac{2 \cdot 76 \times 100}{3 \cdot 83} = 72 \cdot 1$$

Example

The index number of a commodity in 1970 is 152 taking 1960 as base. Calculate, to the nearest whole number, the 1960 index number taking 1970 as base.

1960	1970
x	100
100	152

$$\frac{x}{100} = \frac{100}{152}$$

$$x = 66.$$

Exercise 14.1

1. Find, to the nearest whole number, the index figure for the following commodities **A** to **J**
   (*a*) taking the first value as base,
   (*b*) taking the second value as base.

Commodity	A	B	C	D
First value	£2	£421·85	150 cm	152 kg
Second value	£3	£582	110 cm	160·25 kg

Commodity	E	F	G	H
First value	2300 kg	3·3 m	£1·17½	£0·87
Second value	1970 kg	2·24 m	£4·39	£0·82

Commodity	I	J
First value	1171·13 kg	4 hr 2 min
Second value	1159·37 kg	5 hr 18 min

2. The following table gives the index numbers of five commodities, **A** to **E**, in 1976 taking 1966 as base. Calculate, to the nearest whole number, the 1966 index numbers referred to 1976 as base.

Commodity	**A**	**B**	**C**	**D**	**E**
1966	100	100	100	100	100
1976	112	105	107	123	172

Trends

Relatives are convenient for showing the trend in a single item. Thus the wages in £ million paid in an industry from 1968 to 1976, and the corresponding index number with 1968 as base, are given in the following table.

Year	1968	1969	1970	1971	1972	1973	1974	1975	1976
Wages (£ Million)	233	240	235	246	251	259	265	272	281
Index Number	100	103	101	105	108	111	114	117	121

Other Bases

Other bases may be chosen. Thus if we wished to show how much each item varied from the average, the average amount paid in wages over the whole period is chosen as base; or, if a comparison with the immediate past is required, the wages of any one year can be calculated as a percentage of those paid in the previous year. This is known as the 'chain base' method. Thus using the chain base method the index numbers for the same industry are:

Year	1968	1969	1970	1971	1972	1973	1974	1975	1976
Wages (£ Million)	233	240	235	246	251	259	265	272	281
Index Number	100	103	98	105	102	103	102	103	103

Example

The following index numbers are calculated on the chain base method. Calculate new index numbers if 1973 is taken as base for all the index numbers.

Year	1973	1974	1975	1976
Index Number	100	102	101	105

If the article cost £100 in 1973, it cost £102 in 1974.
In 1975 it cost 101% of its value in 1974, i.e. £103 to the nearest £.
In 1976 it cost 105% of its value in 1975, i.e. £108.
    Therefore the index numbers, taking 1967 as base, are

Year	1973	1974	1975	1976
Index Number	100	102	103	108

Example

In the above example, calculate new index numbers if 1976 is taken as base.

The value of the commodity in 1976 is 105% of its value in 1975. Therefore the work can be arranged as follows:

1975	1976
100	105
x	100

$$x = \frac{100 \times 100}{105} = 95$$

The value of the commodity in 1975 is 101% of its value in 1974.

1974	1975
100	101
x	95

Therefore $x = 94$.

The value of the commodity in 1974 is 102% of its value in 1973.

1973	1974
100	102
x	94

Therefore $x = 92$.

Therefore the new index numbers are

Year	1973	1974	1975	1976
Index Number	92	94	95	100

Exercise 14.2

1. Calculate the price relative to the nearest integer, for the following commodities **A, B, C**, taking the average price for the period as base.

Year		1974	1975	1976	1977	1978
Commodity	**A**	£132	£140	£141	£138	£139
	B	£126	£176	£132	£185	£106
	C	£261	£265	£204	£272	£288

2. The following sets of index numbers are calculated on the chain base method. Calculate, to the nearest whole number, the index numbers (*a*) if 1974 is taken as base, (*b*) if 1978 is taken as base.

Year		1974	1975	1976	1977	1978
Commodity	**A**	100	120	139	142	148
	B	97	102	105	106	110
	C	105	103	100	97	99

Composite Index

The examples so far show the variation in a single item, but, at times, it is necessary to compose an index involving several items. One difficulty is the difference in units in which the items are measured. For example, an index may include the items tobacco and alcoholic drink, the price of tobacco being quoted as so many pence per kilogramme, and drink as so many pence per litre. To overcome the difficulty, the cost of any particular item is replaced by its price relative, which is a pure number.

Suppose a housewife wishes to calculate an index to compare the cost of vegetables in 1979 with the cost in 1970. Her first problem is to decide what vegetables are to be selected to compose the index number. The vegetables chosen must be the ones she is in the habit of buying. Her second problem is that the price of vegetables varies over the year, and so an average or representative price must be decided on.

The following table gives the housewife's choice, the prices decided on, and the price relative:

Vegetable	Price per kg in pence 1970	Price per kg in pence 1979	Price Relative 1970 = 100
Potatoes	5	13	260
Cabbages	7	18	257
Onions	9	20	222
Carrots	12	25	208
Tomatoes	32	50	156
Lettuces	7 each	12 each	171

The housewife buys a larger quantity of potatoes than onions, and it would be unfair for an increase of, say, 10 per cent in the price of onions to affect the index as much as an increase of 10 per cent in the price of potatoes. So she must decide on the relative importance to be attached to each item; that is, each item must be weighted. One method is to weight each item in proportion to the average amount of money spent on it per week. There is no need for great accuracy in the weighting of the items, because slight variations in the weights have little effect on the final index. The weights decided upon were, potatoes 35, cabbage 20, onions 3, carrots 4, tomatoes 29, lettuce 9.

The problem now resolves itself into finding an average. Either the arithmetic, geometric, or any other average may be used. The arithmetic mean is generally the easiest to calculate, while the geometric average is not so easily affected by large values of individual items. In most cases the arithmetic mean is used, the weights replacing the frequencies. The work is set out below.

173

	Price Relative Average prices	Weight	Weight × Price Relative
Vegetable	1970 = 100		
Potatoes	260	35	9100
Cabbages	257	20	5140
Onions	222	3	666
Carrots	208	4	832
Tomatoes	156	29	4524
Lettuces	171	9	1539
Totals		100	21801

Therefore the index number $= \dfrac{21\,801}{100} = 218$.

That is, if a housewife had paid fifty pence in 1970 for a basket of vegetables, in 1979 she would have had to pay $\dfrac{50 \times 128}{100}$, or £1·09 for a similar basket of vegetables.

Index of Retail Prices
A similar index of great importance is the index of retail prices. It has replaced the 'cost of living index' figure. Between June 1947 and January 1956 an 'interim' index was in use, and in 1953–4 an enquiry was made into post-war household spending which provided the information for a new index, which replaced the interim index from January 1956. Details are given in the Ministry of Labour booklet, *Method of Construction and Calculation of the Index of Retail Prices* published by Her Majesty's Stationery Office. In particular, a new set of weights was introduced, based on consumption in 1953–4, valued at 1956 prices. Between January 1962 and 1974, the weights were revised each January on the basis of ascertained consumption in the three years which ended in the previous June, valued at prices current at the date of revision. From 1975 the weights have been revised on the basis of the expenditure for the latest available year. The base of the index is now 15th January 1974 = 100.

Example
Calculate the Index of Retail Prices for December 1975 taking 145 as the working mean. (1974 = 100.)

Article	Index	Weight	Deviation from 145	Weight × Deviation	
Food	144·2	232	− 0·8		− 185·6
Alcoholic Drink	146·6	82	+ 1·6	131·2	
Tobacco	162·2	46	+17·2	791·2	
Housing	134·2	108	−10·8		−1166·4
Fuel and Light	166·8	53	+21·8	1155·4	
Durable Household Goods	141·3	70	− 3·7		− 259
Clothing and Footwear	131·4	89	−13·6		−1210·4
Transport and Vehicles	156·0	149	+11·0	1639	
Miscellaneous Goods	149·1	71	+ 4·1	291·1	
Services	152·5	52	+ 7·5	390	
Meals bought and consumed outside the home	143·6	48	− 1·4		− 67·2
		1000		4397·9	−2888·6

$$\text{Total Index} = 145 + \frac{1509·3}{1000} = 146·5$$

$$= 1509·3$$

Example

A pupil obtains the following percentage marks in an examination. Mathematics 70, French 62, History 65, English 55, Physics 74, Geography 63, Chemistry 67, Machine drawing 82.

It is agreed that the Mathematics, English, and French marks shall have treble weight, and the Physics and Chemistry marks double weight. Calculate the pupil's weighted mean percentage.

Subject	Percentage Mark	Weight	Weight × Percentage Mark
Mathematics	70	3	210
French	62	3	186
History	65	1	65
English	55	3	165
Physics	74	2	148
Geography	63	1	63
Chemistry	67	2	134
Machine Drawing	82	1	82
Totals		16	1053

$$\text{Weighted Mean Percentage} = \frac{1053}{16} = 65\cdot8.$$

Notes on the Index of Retail Prices

The Index of Retail Prices consists of a list of about 350 items as diverse as the price of bread, the cost of watch repairs, car bills and cinema tickets. The changes in the prices of each item are obtained from a great number of sources, and the weights to be allotted to each are mainly obtained from the Family Expenditure Survey. The troubles start when the figures are used as a basis for wage claims or for increased grants or pensions. The Department of Employment and Productivity points out that the index is not a true measure of the cost of living, which is why the representative body which oversees the compilation of the index is now called the Retail Price Index Advisory Committee.

The index breaks down as a cost of living index because it does not include a number of items which are important expenses to most families—items such as Insurance premiums, national savings stamps and tax. Another drawback is the difference in the importance of various items to a family depending on where it lives and its income. An increase in rail fares will matter far more to a commuter than to someone living in the middle of London, and a rise in the cost of food affects a family of mother and father and five children with an income of £60 a week far more than a family of husband and wife and a combined income of £140 a week.

Many wage agreements are tied to a measure of the cost of living. Official objection to these agreements is on inflationary grounds, while the unions object to the ways in which the index underates the actual real rise in costs. Thus the index fails to take taxation into account. Tax appreciably reduces the value of every £1 awarded in wages, but every £1 in the rise of the cost of living is real. Another problem is the cost of buying a house. A person owning his own house is assessed on the basis of rentable value and this does not reflect increases in the cost of mortgages or the declining standard of housing at any given price.

Exercise 14.3

1. Calculate a composite index number from the following index numbers weighted as shown:

   (a)
Index Number	102	132	140	120	112
Weight	1	3	4	6	2

   (b)
Index Number	134	152	196	92	86
Weight	2	4	3	5	1

(c)

Index Number	47	53	84	91	73
Weight	3	3	2	2	1

2. Calculate from the following table an index of retail prices. Give your answer correct to the nearest integer.

Item	Price Relative	Weight
Food	122	348
Rent	101	88
Clothing	118	95
Fuel	115	67
Household Goods	111	73
Miscellaneous	108	100

3. Calculate from the following table an index of retail prices. Take 110 as a working mean, and give your answer correct to the nearest integer.

Item	Price Relative	Weight
Food	114	400
Rent	112	70
Clothing	96	100
Fuel	110	70
Household Goods	95	60
Miscellaneous	108	100

Common Index Figures

Other common index figures are defined as follows:

Birth rates are expressed as so many live-births per thousand of the estimated population at the middle of the year. Thus in the U.K. in 1967 the number of births was 616600 and the estimated population was 55202000. Therefore the birth rate was $\dfrac{616600 \times 1000}{55202000} = 11\cdot2$.

Death rates are expressed as the number of deaths occurring every year per 1000 of the estimated population at the middle of the year.

Survival rate is the difference between the birth rate and the death rate.

Marriage rate is the number of persons marrying per thousand of the population at the middle of the year.

Infant mortality rate is the death rate of children under one year of age per thousand live births.

The above rates are known as crude rates and may give a false picture of the true state of affairs. A better picture is given by a standardised rate.

Standardised or Corrected Rates

For a true comparison of two rates, more information is required. Thus in two towns **A** and **B**, there are in one year 93 and 37 deaths, respectively, and the population of each town is 5000. Therefore the crude death rate for town **A** is $\dfrac{93 \times 1000}{5000}$, that is 18·6; and for town **B**, $\dfrac{37 \times 1000}{5000}$, that is 7·4.

Anyone comparing these rates may assume that town **B** is healthier than town **A**. On further analysis the populations and deaths are shown to be:

Age Group	Population Town **A**	Population Town **B**	Deaths Town **A**	Deaths Town **B**
0 and under 20	1500	2500	9	15
20 and under 40	2000	2000	4	4
40 and under 60	1000	400	10	4
60 and over	500	100	70	14
Totals	5000	5000	93	37

If the crude death rate is calculated in each age group for each town, we have:

Age Group	Death Rate Town **A**	Death Rate Town **B**
0 and under 20	6	6
20 and under 40	2	2
40 and under 60	10	10
60 and over	140	140

Now, since the death rates in both towns at each age group are equal, it is only reasonable to expect the final death rates to be equal. The difference in crude death rate is due to the difference in numbers in the age groups. To give a corrected or standardised death rate, the death rate in each group, known as the specific death rate, is weighted according to the number of persons in the *country* in that age group.

Thus in England and Wales the number of persons in each age group to the nearest million is:

0 and under 20	15 million
20 and under 40	12 million
40 and under 60	12 million
60 and over	9 million
Total	48 million

Therefore the corrected death rate for each town is:

$$\frac{15 \times 6 + 12 \times 2 + 12 \times 10 + 9 \times 140}{48} = 31 \cdot 1.$$

Age distributions other than the one named above may be used. Since the census statistics give the total population of the country divided into age groups of 0 and under 4 years, 4 years and under 9 years, 9 and under 15 years, and so on, it is general to take this grouping.

If $p_1, p_2, \ldots p_n$, are the populations, and $d_1, d_2, \ldots d_n$ the death rates of the various age groups, and P is the total population, i.e. $P = p_1 + p_2 + \ldots + p_n$, then the corrected death rate is

$$\frac{p_1 d_1 + p_2 d_2 + \ldots + p_n d_n}{P}$$

This can be written as

$$\frac{1}{100}\left[\frac{100 p_1}{P} d_1 + \frac{100 p_2}{P} d_2 + \ldots + \frac{100 p_n d_n}{P}\right]$$

Now $\dfrac{100 p_1}{P}$ is the percentage population in the first age group, and $\dfrac{100 p_2}{P}$ is the percentage population in the second age group, and so on.

Let $\dfrac{100 p_1}{P} = R_1, \dfrac{100 p_2}{P} = R_2$, and so on.

Then the corrected death rate is

$$\frac{1}{100}[R_1 d_1 + R_2 d_2 + \ldots + R_n d_n]$$

Example

The following table gives the death rates in age groups for a certain town in the United Kingdom for last year, and the percentage population of the United Kingdom in each age group. Calculate the standardised death rate for the town.

Age Group	0–4	5–14	15–24	25–34	35–44	45–54	55–64	65 and over
Death Rate	6·0	0·4	1·5	1·3	2·7	7·9	22·2	83·4
% of Population	7	16	13	15	12	14	13	10

From the previous paragraph, the standardised death rate is

$$\frac{1}{100}[7 \times 6·0 + 16 \times 0·4 + 13 \times 1·5 + 15 \times 1·3 + 12 \times 2·7 + 14 \times 7·9$$
$$+ 13 \times 22·2 + 10 \times 83·4] = 13·5.$$

A similar method of correction is applicable in comparing birth rates, marriage rates, or any other attributes, when it is necessary to refer the attributes to a standard population to avoid complications due to different age distributions.

Exercise 14.4

1. In each of the following tables calculate an index figure for each year by taking (i) the first year as base, (ii) the last year as base, (iii) the arithmetic mean of the numbers for the period as base, (iv) the first year as 100 and using the chain base method.

   (*a*) Attendance at cinemas in Great Britain.

Year	1972	1973	1974	1975	1976
Admissions (M)	157	134	138	116	104

   (*b*) Number of pupils in maintained primary and secondary schools in England and Wales.

Year	1967	1968	1969	1970	1971	1972	1973
Pupils (Thousand)	7328	7542	7753	7960	8167	8366	8514

Year	1974	1975	1976
Pupils (Thousand)	8873	8926	8984

2. The following index figures are calculated on the chain base method. Convert them into index figures (*a*) using the year 1970 as base, (*b*) using the year 1978 as base.

Year	1970	1971	1972	1973	1974	1975	1976	1977	1978
Index	100	103	98	101	104	102	101	103	104

3. The index figures of four commodities **A, B, C, D**, in 1978 taking 1964 as base, and the weights allotted to them in a composite index are:

Commodity	1964	1978	Weights
A	100	176	4
B	100	120	1
C	100	264	2
D	100	296	3

   Calculate the composite index figure.
   Repeat the calculation using the weights 5, 2, 3, 4, respectively. Is there any conclusion to be drawn?

4. A composite index number is to be constructed in each of the following cases (*a*) to (*e*) using the index numbers and weights shown. Select a suitable working mean and calculate the composite index number as an arithmetic mean.

   (*a*)
Index Number	184	176	142	129	136	
Weight	1	2	3	4	5	

   (*b*)
Index Number	122	109	193	191	182	132
Weight	2	4	7	1	7	2

   (*c*)
Index Number	188	198	123	199	164	
Weight	5	8	7	7	2	

(d) Index Number	129	106	166	114	171		
Weight	2	4	3	9	6		
(e) Index Number	169	122	180	171	127	175	187
Weight	6	8	4	7	3	1	8

5. A school is biased towards the study of Mathematics and Science, and, in calculating a pupil's final mark, allots weights to the percentage marks in each subject.

The following table gives the marks of seven pupils and the weights allotted to each subject. Calculate the pupils' final marks.

		Pupils						
Subject	Weights	A	B	C	D	E	F	G
Mathematics	6	56	40	57	32	51	47	71
English	4	74	39	52	26	84	40	66
Languages	2	62	41	51	29	63	42	63
Science	6	69	51	62	40	51	51	73
History	1	57	54	60	27	62	74	40
Geography	1	48	53	47	35	56	82	39
Art	1	62	61	63	57	54	71	36
Handicrafts	3	59	60	59	43	57	52	64

6. The following table gives the male and female population in age groups in the United Kingdom for a year, and the number of male and female deaths in each group during the year. Calculate the male and female death rates for each age group.

Age Group	Males		Females	
	Population (Thousand)	Deaths	Population (Thousand)	Deaths
0–4	1980	13281	1883	9689
5–9	2189	998	2091	695
10–14	1897	778	1814	502
15–19	1675	1277	1630	586
20–24	1665	1661	1621	878
25–34	3563	4385	3561	3201
35–44	3555	9037	3664	7123
45–54	3545	26997	3705	17148
55–64	2493	55719	3065	35320
65–74	1566	85540	2232	72565
75–84	661	84310	1093	99828
85 and over.	90	23215	192	43428

(A.)

7. With the help of your parents, construct an index figure to compare the cost of vegetables used in your home today with their cost ten years ago.

8. A composite index number is to be constructed from the following index numbers weighted as shown. Using a working mean of 140, or otherwise, calculate the composite index number as an arithmetic mean.

Index Numbers	172	166	150	135	130	
Weights	1	2	4	6	3	(L)

9. What is meant by 'weighting' in the construction of index numbers? Using a working mean of 115, or otherwise, calculate as an arithmetic mean correct to the nearest whole number the 'interim index of retail prices', using the following table of price relatives and weights.

	Price Relatives	Weights	
Food	122·0	348	
Rent and Rates	101·3	88	
Clothing	118·4	97	
Fuel and Light	115·2	65	
Household durable goods	110·6	71	
Miscellaneous goods	113·3	35	
Services	106·6	79	
Drink and tobacco	107·5	217	(L)

10. Explain what is meant by (i) crude death rates, (ii) corrected death rates.

The following table gives the population by age groups of a certain town for the year 1952 and the number of deaths occurring in each age group during the year. Calculate correct to one decimal place (a) the crude death rate, (b) the death rate in each age group, (c) the corrected death rate.

Age group	0–5	5–15	15–25	25–35	35–45	45–55	55–65	over 65
Population	1000	1250	600	900	800	750	500	400
Deaths in each age group	13	1	2	3	3	8	12	20
% of total population of Great Britain in each age group	12·1	17·7	15·8	15·1	14·0	11·6	7·7	6·0

(L)

11. Explain what is meant by an index number

Calculate as an arithmetic mean correct to the nearest integer, a cost of living index from the following table of price relatives and weights.

	Price Relative	Weight
Food	122	35
Rent	101	9
Clothing	118	10
Fuel	115	7
Miscellaneous	108	39

(L)

12. In 1952 the index of shipping freight based on 1948 as 100 was 110·6. In 1954 the index based on 1952 as 100 was 86·1. What would have been the index in 1954 based on 1948 as 100? (L)

13. Contrast the crude death rate with the standardised death rate. Find the difference between the standardised death rates for two towns, **A** and **B**, from the following data.

Age Group	Death rates per thousand **A**	**B**	Percentage of total population of Great Britain
0–5	14	13	12·1
5–15	2	1	17·7
15–25	2	3	15·8
25–35	4	3	15·1
35–45	5	6	14·0
45–55	9	10	11·6
55–65	18	22	7·7
Over 65	40	63	6·0

(L)

14. The cost of living index is calculated as a weighted mean. From the following table calculate the rise in this index between 1970 and 1974.

	Weight	Index of Retail Prices (1962 = 100) 1970	1974
Food	253	140·1	230·0
Alcoholic Drink	70	143·9	182·1
Tobacco	43	136·3	164·8
Housing	124	158·1	238·2
Fuel and Light	52	145·7	208·8
Durables	64	126·0	170·8
Clothing and Footwear	91	123·8	182·3
Transport and Vehicles	135	132·1	194·3
Miscellaneous	63	142·8	202·7
Services	54	153·8	227·2
Meals bought outside	51	145·5	248·3

SOURCE: Dept. of Employment

183

15. Calculate, as a weighted mean correct to one decimal place, the total index of industrial production in January 1975 from the following table.

	Weight	Index (Average 1970 = 100)
Total manufacturing	745	101·4
Mining and quarrying	37	86·3
Construction	146	93·3
Gas, Electricity and Water	72	120·4

SOURCE: C.S.O.

If the index for gas and electricity is omitted and the other data remain the same, find the change in the total index of industrial production.

16. The 1958 index of industrial production for the Engineering industries was 112, referred to 1954 as base. Calculate the 1954 index referred to 1958 as base. (L)

17. Calculate the standardised death rate from the following data:

Age group	0–5	5–15	15–25	25–35	35–45	45–55	55–65	over 65
Death rates per thousand	11	3	2	2	7	12	18	58
% of total population of Great Britain	12·1	17·7	15·8	15·1	14·0	11·6	7·7	6·0

(L)

18. Some price relatives for food are given in the table below:

	Price relatives	
	June 1972 (Average 1963 = 100)	August 1976 (June 1972 = 100)
Beef	120	161
Mutton and lamb	110	166
Bacon and ham	120	152
Butter	100	188
Margarine, cooking fats	160	144
Cheese	92	140
Eggs	92	201
Tea	125	122
Sugar and Syrup	125	186

(*a*) Using the geometric mean, construct a simple unweighted index number for August 1976, taking 1972 as 100.

(*b*) Construct price relatives for August 1976 taking the 1963 average as 100, and deduce the corresponding value of a simple unweighted index number, using the arithmetic mean.

19.   Population and deaths of infants under five years old:
(England and Wales)

	Under 2 years old		2 years and under 5 years		(Thousands) Total	
Year	Population	Deaths	Population	Deaths	Population	Deaths
1931	1195	48·9	1796	8·4	2991	57·3
1951	1367	22·0	2355	2·4	3722	24·4

Calculate crude death rates for the infant population in both years. Calculate age-specific death rates, and hence the standardised death rate for 1951, taking the infant population of 1931 as standard.

Describe in words the most prominent features of the data revealed by your analysis. (*J.M.B.*)

20.   The first three rows of the following table show extracts from the vital statistics of a certain town while the last row gives an assumed standard age distribution per thousand:

Age group	0–9	10–39	40–59	60–
Population in thousands	16	43	12	4
Deaths in 1955	212	247	291	337
Standard age distribution	150	400	300	150

Calculate the crude and standardised death rates and explain why crude death rates are considered inadequate. (*J.M.B.*)

21. (*a*) The table gives the price relative for a commodity for the four years 1975–78. The price relative is calculated on the chain-base method, that is, the value of the item is calculated as a percentage of its value in the previous year.

Year	1975	1976	1977	1978
Price Relative (Chain base)	97·5	103	98	102

If the commodity cost £12·5 in 1974, calculate its value in each of the years 1975 to 1978.

(*b*) The 1966 and 1971 index numbers for volume of imports were 110 and 135 (1964 = 100).

(i) What are the 1964 and 1971 index numbers if 1966 = 100?

(ii) What are the 1964 and 1966 index numbers if 1971 = 100?

22. (a) Explain what is meant by crude death rates and corrected death rates.
    (b) The table gives the population in age groups of a certain town for the year 1977 and the number of deaths occurring in each age group during the year.

    Calculate (i) the crude death rate,
    (ii) the corrected death rate for the town.

Age Group (years)	Population	Deaths in each Age Group	Percentage of the total population in each age group
0–15	2500	15	29·4
15–35	2000	7	31·2
35–55	1500	13	26·2
55–75	900	32	10·1
Over 75	150	20	3·1

(A.E.B.)

23. The following index numbers are calculated on the chain base method. Calculate the new set of index numbers taking 1974 = 100.

Year	1974	1975	1976	1977
Index Numbers (Wages)	101	104	104	107

24. (a) Define 'weighted mean' and state its advantages over a simple mean.
    (b) On a certain date, the index of retail prices was 109·1. Percentage increases over the base year and the weights allocated to each group are given in the table. What was the percentage increase in the Food Group?

Group	Percentage Increase	Weight
Food		400
Rent	12	70
Clothing	−4	100
Fuel	10	70
Household Goods	−5	60
Miscellaneous	8·4	100
		800

(A.E.B.)

15 Dispersion

Dispersion

An average was defined as a number that summarises all or some of the characteristics of a distribution. In many cases an average gives sufficient information, but in other cases more information may be required. For example, consider the scores of two boy cricketers, **A** and **B**, as given in the following table:

Runs scored   1, 2, 3, 4, 5, 6, 7, 8, 9, 10.
Number of **A**   0, 1, 1, 1, 6, 8, 6, 1, 1, 1.
Innings   **B**   2, 2, 1, 3, 3, 3, 3, 3, 3, 3.

In each case the total number of runs scored is 156, and the total number of innings played is 26, so that the arithmetic mean in each case is 6. But the distributions are entirely different, as is evident if histograms are drawn.

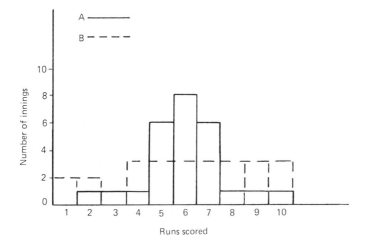

Cricketer **A**'s scores are concentrated around the arithmetic mean, but cricketer **B**'s scores are widely dispersed. It is therefore useful to have an index to indicate whether the items of a distribution are closely clustered around the average, or whether they are spread over the range of the distribution, and to what extent.

The scatter of the items of a distribution is known as dispersion, and the index giving the measure of the dispersion is known as the coefficient of deviation.

Mean Deviation

One method of measuring dispersion is to find the deviations of all the items from the average, ignore their signs, and find the arithmetic mean of their magnitude. This is known as the *mean deviation*. The deviation of each item can be measured from the mean, mode, median, or any other value chosen, but it is usual to measure it from the arithmetic mean of the distribution. Thus in the case of cricketer **A** we have:

Runs scored	Number of Innings	Deviation from the mean, 6	
	f	d	fd
1	0	-5	0
2	1	-4	-4
3	1	-3	-3
4	1	-2	-2
5	6	-1	-6
6	8	0	0
7	6	1	6
8	1	2	2
9	1	3	3
10	1	4	4
Totals	26	(Ignoring signs)→	30

Mean Deviation $= \dfrac{30}{26} = 1\cdot15$ (2·31 in the case of cricketer **B**).

Note 1 If the signs of the deviations about the arithmetic mean are not neglected their sum is zero. (See Theorem 1, page 191.)

Example

Find the mean deviation from the arithmetic mean of 33, 39, 27, 44, 36, 40, 55, 37, 50, 39.

The sum of the numbers is 400. The arithmetic mean is $\dfrac{400}{10} = 40$.

Deviation from the mean -7, -1, -13, 4, -4, 0, 15, -3, 10, -1. The sum of the deviations ignoring signs is 58.

Therefore the mean deviation $= \dfrac{58}{10} = 5\cdot8$.

Exercise 15.1

Find the mean deviation from the arithmetic mean of the following sets of numbers, and in each case verify that if the signs of the deviations are not neglected their sum is zero.

1. 12, 17, 13, 18, 10, 16, 14, 12, 18, 10.
2. 26, 25, 20, 27, 29, 28, 21, 22, 23, 29.
3. 41, 47, 32, 46, 46, 31, 38, 34, 38, 49, 43, 33.
4. 41, 58, 51, 51, 51, 43, 53, 54, 46, 57, 53, 51, 55, 54.

Root-Mean-Square Deviation and Standard Deviation

Another method of calculating dispersion is to find the square-root of the average of the squares of the deviations from a working mean. This is called the root-mean-square deviation, but if the deviations are measured from the arithmetic mean of the distribution, it is called the standard deviation.

The small Greek letter σ, called sigma, is the symbol used to denote standard deviation, and the letter s is used to denote the root-mean-square deviation.

Thus in the case of batsman **A**:

Runs scored	Number of Innings	Deviations from mean		
	f	d	fd	fd^2
1	0	-5	0	0
2	1	-4	-4	16
3	1	-3	-3	9
4	1	-2	-2	4
5	6	-1	-6	6
6	8	0	0	0
7	6	1	6	6
8	1	2	2	4
9	1	3	3	9
10	1	4	4	16
Totals	26		0	70

$$\text{Standard deviation, } \sigma = \sqrt{\frac{70}{26}} = 1 \cdot 64.$$

In the case of batsman **B** the standard deviation is $2 \cdot 75$.

Example

Find a the standard deviation, b the root-mean-square deviation from a working mean of 10 for the following numbers: 19, 17, 19, 10, 11, 18, 14, 19, 11, 12.

(a) Standard deviation.

											Total
Numbers	19	17	19	10	11	18	14	19	11	12	150
Deviation from mean $\left(\dfrac{150}{10} = 15\right)$	4	2	4	−5	−4	3	−1	4	−4	−3	
Deviation squared	16	4	16	25	16	9	1	16	16	9	128

$$\sigma^2 = \frac{128}{10} = 12\cdot8$$
$$\sigma = \sqrt{12\cdot8} = 3\cdot6$$

(b) Root-mean-square deviation

											Total
Numbers	19	17	19	10	11	18	14	19	11	12	
Deviations from working mean of 10	9	7	9	0	1	8	4	9	1	2	
Deviation squared	81	49	81	0	1	64	16	81	1	4	378

$$s^2 = \frac{378}{10} = 37\cdot8$$
$$s = \sqrt{37\cdot8} = 6\cdot1$$

Note 2 If b denotes the difference between the value of the working mean and the arithmetic mean (in this example $b = 5$) we have

$$\sigma^2 + b^2 = 12\cdot8 + 25 = 37\cdot8 = s^2$$

See Theorem 2, page 191.

Exercise 15.2
In the following sets of numbers find *a* the standard deviation, *b* the root-mean-square deviation from the working means given in the brackets; in each case test if $s^2 = \sigma^2 + b^2$.

(a) 13, 12, 14, 15, 16, 17, 18, 10, 16, 19.   (10, 12, 14, 17.)
(b) 21, 26, 27, 29, 22, 22, 27, 26.   (27, 29, 31.)
(c) 130, 126, 149, 150, 180.   (140, 142, 150, 155.)
(d) 17, 13, 12, 8, 17, 6, 11, 11, 5, 10.   (12, 4, 0, 16.)
(e) 34, 35, 32, 34, 32, 37, 34.   (24, 30, 36.)

Theorems
The two statements made in Notes 1 and 2 are very important, and are used to find the standard deviation of a frequency distribution. In the above exercises we verify the results in particular cases; we will now prove them generally.

Theorem 1

In a set of numbers, the sum of the deviations of the individual items from their arithmetic mean is zero.

Let the items be $x_1, x_2, \ldots x_n$, and M their arithmetic mean.

Since there are n items

$$x_1 + x_2 + \ldots x_n = nM$$

The deviations from the arithmetic mean are

$$(x_1 - M), (x_2 - M), \ldots (x_n - M)$$

and the sum of the deviations is

$$(x_1 - M) + (x_2 - M) + \ldots (x_n - M) = x_1 + x_2 + \ldots x_n - nM = 0$$

Theorem 2

The square of the root-mean-square deviation is equal to the square of the standard deviation together with the square of the difference between the arithmetic mean and the working mean.

Let the values of the items be $x_1, x_2, \ldots x_n$, the working mean W and the arithmetic mean M.

Let $W - M = b$, i.e. $W = b + M$.

The deviations from the working mean are:

$$(x_1 - W), (x_2 - W), \ldots (x_n - W)$$

and the deviations from the arithmetic mean are:

$$(x_1 - M), (x_2 - M), \ldots (x_n - M)$$

The square of the root-mean-square deviation is

$$\frac{1}{n}[(x_1 - W)^2 + (x_2 - W)^2 + \ldots (x_n - W)^2]$$

$$= \frac{1}{n}[(x_1 - M - b)^2 + (x_2 - M - b)^2 + \ldots (x_n - M - b)^2]$$

$$= \frac{1}{n}[(x_1 - M)^2 - 2b(x_1 - M) + b^2 + (x_2 - M)^2 - 2b(x_2 - M) + b^2 \ldots$$

$$+ (x_n - M)^2 - 2b(x_n - M) + b^2]$$

$$= \frac{1}{n}[(x_1 - M)^2 + (x_2 - M)^2 + \ldots (x_n - M)^2$$

$$- 2b\{(x_1 - M) + (x_2 - M) + \ldots (x_n - M)\} + nb^2]$$

But $\frac{1}{n}[(x_1 - M)^2 + (x_2 - M)^2 + \ldots (x_n - M)^2]$ is by definition equal to σ^2, and $(x_1 - M) + (x_2 - M) + \ldots (x_n - M)$ is the sum of the deviations from the arithmetic mean, and this, by Theorem 1, equals zero. Therefore,

$$s^2 = \sigma^2 + b^2$$

Corollary: s^2 is least when $b^2 = 0$; i.e. the standard deviation is the least root-mean-square deviation of a distribution.

Exercise 15.3

1. Fill in the blanks in the following table:

s	σ	b
5		3
3	2	
	3	1
3·7	3·2	
4·2		2·1
	2·7	1·5

2. The arithmetic mean and the standard deviation of a distribution are 33·2 and 2·5 units, respectively. What is the root-mean-square deviation from 36·4 units?

σ for a Frequency Distribution

The relation $s^2 = \sigma^2 + b^2$ simplifies the calculation of the standard deviation of a frequency distribution.

Example

Find the standard deviation of the following set of marks.

Make the same assumption as in averages, that is, all the candidates in any one class have the same marks—namely the marks at the middle of the class-interval.

Col. 1	Col. 2	Col. 3	Col. 4	Col. 5	Col. 6
			d = deviation	fd in units	fd^2 in units
	Middle	f	from working	of 10	of 10^2
Class	Mark	Frequency	mean in units		
			of 10		
1–10	5·5	5	−4	−20	80
11–20	15·5	9	−3	−27	81
21–30	25·5	16	−2	−32	64
31–40	35·5	30	−1	−30	30
				−109	
41–50	45·5	56	0	0	0
51–60	55·5	45	1	45	45
61–70	65·5	25	2	50	100
71–80	75·5	10	3	30	90
81–90	85·5	2	4	8	32
91–100	95·5	2	5	10	50
Total		200		143	572
				−109	
				34	

The arithmetic mean = working mean + average of the deviations from the working mean

$$= 45 \cdot 5 + \frac{10 \times 34}{200}$$

$$= 47 \cdot 2$$

$$b = \frac{10 \times 34}{200} = 1 \cdot 7$$

$$b^2 = 2 \cdot 89$$

$$s^2 = \frac{572}{200} = 2 \cdot 86 \text{ in units of } 10^2$$

$$= 286$$

$$\sigma^2 = s^2 - b^2$$

$$= 286 - 2 \cdot 89$$

$$= 283$$

$$\sigma = 16 \cdot 8 \text{ marks}$$

$$= 17 \text{ marks, to two significant figures.}$$

Method: Col. 1 is the class, Col. 2 the middle mark, and Col. 3 the frequency. Col. 4 is the deviation from the working mean, 45·5, in units of 10 marks. By picking units of 10 we have smaller numbers to deal with.

Col. 5 is obtained by multiplying Col. 3 by Col. 4, and is therefore in units of 10 marks.

Col. 6 is obtained by multiplying Col. 5 by Col. 4, and is therefore in units of 10 marks squared.

Meaning of the Coefficient of Standard Deviation

It is essential to understand what the coefficient of standard deviation measures, and the following explanation may help to make it clear.

If we pick at random an item out of a distribution, the standard deviation gives us a measure of the unlikelihood of the value of that item being near the arithmetic mean of the distribution. The greater the measure of deviation, the greater the chance of missing the arithmetic mean.

If the distribution is fairly symmetrical and bell-shaped and the number of items is large, i.e. what is known as a normal distribution, it can be shown that approximately 0·68 of the items lie in the range $M \pm \sigma$, 0·95 of the items in the range $M \pm 2\sigma$, and nearly all the items in the range $M \pm 3\sigma$.

Thus in the example of dispersion on page 192, the arithmetic mean was approximately 47 and the standard deviation 17.

Therefore

$M \pm \sigma = 47 \pm 17$ or 30 to 64, and proportion of items in this range
$$= \frac{144}{200} \text{ or } 0 \cdot 72$$

$M \pm 2\sigma = 47 \pm 34$ or 13 to 81, and proportion of items in this range
$$= \frac{189}{200} \text{ or } 0 \cdot 945$$

$M \pm 3\sigma = 47 \pm 51$ or -4 to 98, and proportion of items in this range
$$= \frac{200}{200} \text{ or } 1$$

These agree very closely with the theoretical values given above. At times the standard deviation is known as the standard error, and the square of the standard deviation as the variance.

Important Application of Theorem 2

An important application of Theorem 2 is when the working mean is zero. The root-mean-square deviation from zero is then

$$\frac{1}{n}(x_1^2 + x_2^2 + \ldots + x_n^2)$$

where $x_1, x_2, \ldots x_n$ are the values of the variable.

The difference between the values of the working mean and the arithmetic mean is the value of the arithmetic mean.
Therefore,
$$b = \frac{1}{n}(x_1 + x_2 + \ldots + x_n) = \text{A.M.}$$

Therefore,

$$\sigma^2 = \frac{x_1^2 + x_2^2 + \ldots + x_n^2}{n} - \left[\frac{x_1 + x_2 + \ldots + x_n}{n}\right]^2$$

or, in words, *the variance equals the arithmetic mean of the squares of the variables minus the square of the arithmetic mean of the values of the variable.*

Example

Calculate the mean and standard deviation of the following set of numbers: 4, 5, 6, 8, 4, 9.

x : 4   5   6   8   4   9   Total $=$ 36
x^2: 16   25   36   64   16   81   Total $= 238$

A.M. $= \dfrac{36}{6} = 6.$

$\sigma^2 \quad = \dfrac{238}{6} - (6)^2 = 3 \cdot 67. \qquad \sigma = 1 \cdot 92.$

Example

Calculate the arithmetic mean and standard deviation of the following distribution.

x (mid-class):		1	2	3	4	5
f	:	4	8	6	3	1

x	f	fx	fx^2
1	4	4	4
2	8	16	32
3	6	18	54
4	3	12	48
5	1	5	25
Total	22	55	163

$$\text{A.M.} = \frac{55}{22} = 2 \cdot 5.$$

$$\sigma^2 = \frac{163}{22} - (2 \cdot 5)^2 = 1 \cdot 159. \qquad \sigma = 1 \cdot 08.$$

This method becomes cumbersome if the values of the variable are large, but is useful for the solution of many problems.

Example

Given that in a distribution

$$\text{A.M.} = 25 \cdot 3; \ \sigma^2 = 41 \cdot 0; \ n = 321,$$

calculate the sum of the squares of the values of the variable.

$$\sigma^2 = \frac{(x_1^2 + x_2^2 + \ldots + x_n^2)}{n} - M^2$$

Therefore,

$$41 \cdot 0 = \frac{x_1^2 + x_2^2 + \ldots + x_n^2}{321} - (25 \cdot 3)^2$$

and,

$$x_1^2 + x_2^2 + \ldots + x_n^2 = 218\,630$$

Exercise 15.4

1. Calculate (i) the mean deviation, and (ii) the standard deviation for the examples in Exercises 12.2 and 12.3.

   Find the proportion of the number of items in the ranges $M \pm \sigma$, $M \pm 2\sigma$, $M \pm 3\sigma$. If the proportion differs by a large amount from the theoretical values given above, offer an explanation.

2. Find (i) the arithmetic mean (ii) the standard deviation, of the following set of numbers

(*a*) without further grouping,
(*b*) grouping in intervals of five, (1–5, 6–10, etc.)
(*c*) grouping in intervals of ten, (1–10, 11–20, etc.)
(*d*) grouping in intervals of twenty, (1–20, 21–40, etc.)
1, 2,2, 6, 6, 8, 11, 13, 13, 13, 17, 20, 22, 22, 24, 24, 24, 27, 28, 29, 30, 32, 36, 39, 40, 40, 44, 44, 47, 47, 47, 51, 53, 55, 59, 59, 61, 61, 63, 65, 65, 67, 69, 70, 75, 75, 75, 80, 81, 87, 90, 91, 92, 95, 99, 100. (*Note:* Generally the mean is very little affected by the grouping, but the standard deviation increases with an increase in the range of the class-interval.)

3. Find the mean, the mean deviation, and the standard deviation of all integers from 1 to 39. Repeat for the following random sample and compare: 31, 12, 18, 33, 4, 36, 26, 13, 6, 1.

4. The mean height of a normal group of 1000 men is 168 cm, and the standard deviation of the group is 5 cm. State the probable height of the shortest, and the tallest man in the group.

   What is the probable number of men in the following groups: (*a*) 163 cm to 173 cm, (*b*) 158 cm to 178 cm?

5. One normal distribution of 2000 men has a mean mass of 64 kg and a standard deviation of 3 kg, while another normal distribution of 2000 men has a mean mass of 66 kg and a standard deviation of 2 kg. Which do you consider has the heaviest man, and why?

6. A normal distribution of marks has an arithmetic mean of 60 and a standard deviation of 20. If the pass mark is 40, estimate
   (*a*) the percentage of candidates with marks over 80.
   (*b*) the percentage of candidates who pass.
   (*c*) the percentage of candidates with marks over 100.

Other Forms of Dispersion
Other methods of measuring dispersion have been suggested, and one very simple and quick method is to find the difference between the two extreme items of the distribution. This has many disadvantages. In the case of our two batsmen it would give no extra information, as the *range*, that is the difference between the extreme items, is nearly the same in each distribution.

A better measure is the quartile deviation, which is half the difference between the upper and lower quartiles = $(Q_3 - Q_1)/2$. It is also known as the *semi-interquartile range*.

The desirable properties for a coefficient of deviation are the same as those for an average. That is, it should be calculable by all observers.

be quickly calculated, only slightly affected by fluctuations of sampling, and lend itself to algebraical treatment.

The range is not based on all the items, is badly affected by fluctuations of sampling, does not lend itself to algebraical treatment, and its use is not recommended. The quartile deviation is calculated with great ease, and has a clear and simple meaning, but owing to the absence of simple algebraical properties, and the difficulty of its behaviour in sampling its use is not recommended unless the calculation of the standard deviation is difficult.

The mean deviation is rigidly defined, is based on all the observations, and is calculated with ease, but it is the standard deviation that possesses the majority of properties mentioned. It is rigidly defined, is based on all the observations, is calculated with reasonable ease, lends itself to algebraical treatment, and is found to be the index least affected by fluctuations of sampling.

If a rough estimate of the dispersion is required, the quartile deviation is satisfactory if the distribution of the items is fairly normal; but in a moderately skew distribution, the standard deviation is by far the most useful and important measure of dispersion.

The following relations are approximately true in a fairly normal distribution:

$$\text{Quartile deviation} = \frac{2}{3} \text{ standard deviation.}$$

Mean deviation measured from the arithmetic mean

$$= \frac{4}{5} \text{ standard deviation.}$$

Absolute Measures of Dispersion

The measures of dispersion discussed have all been expressed in terms of the units of the items of the distributions. It is thus impossible to compare dispersions in different units. For this reason it has been suggested that 'absolute' measures of dispersion, that is measures that are pure numbers and not expressed in any unit, should be used.

The most common are:

1. Quartile coefficient of dispersion $= \dfrac{Q_3 - Q_1}{Q_3 + Q_1}$

2. Coefficient of mean dispersion

$$= \frac{\text{mean deviation from arithmetic mean}}{\text{arithmetic mean}}$$

$$\text{or} = \frac{\text{mean deviation from median}}{\text{median}}$$

$$\text{or} = \frac{\text{mean deviation from mode}}{\text{mode}}$$

3. Coefficient of dispersion $= \dfrac{\text{standard deviation}}{\text{arithmetic mean}}$

With all coefficients of dispersion, the smaller the coefficient the less the dispersion of the items.

If the coefficient of dispersion is expressed as a percentage, it is known as the coefficient of variation.

$$\text{Coefficient of variation} = \frac{\sigma}{M} \times 100$$

The coefficient of variation is useful for comparing distributions of the same type. For example, if one mass-produced article has a mean length of 30 cm and a standard deviation of 0·6 cm, the coefficient of variation is

$$\frac{0\cdot6}{30} \times 100 = 2$$

If another mass-produced article has a mean length of 10 cm and a standard deviation of 0·15 cm, then the coefficient of variation is

$$\frac{0\cdot15}{10} \times 100 = 1\cdot5$$

It becomes evident at a glance that there is a higher degree of precision of manufacture in the second article. Note that the coefficient of variation is a number and is independent of units. It can be used to compare distributions of the same type but in different units. It becomes unreliable if M tends to zero.

Measurement of Skewness
In the chapter on Frequency Distributions, the term skewness was defined. Many measures have been suggested to give a measurement of skewness, and the most common is known as Pearson's measure:

$$\text{skewness} = \frac{\text{mean} - \text{mode}}{\text{standard deviation}} \qquad (a)$$

This is a pure number and is zero for symmetrical distributions. The drawback of this measure is the difficulty in calculating the mode, but if the empirical relationship given in the chapter on Averages, [viz. mode = mean − 3(mean − median)] is used for the mode, we then have

$$\text{skewness} = \frac{3(\text{mean} - \text{median})}{\text{standard deviation}} \qquad (b)$$

Exercise 15.5

Calculate (i) the range,

(ii) the semi-interquartile range,

(iii) the skewness, (using formula (a) p. 198),

for examples 1 to 5 of Exercise 12.3 on pages 135–136.

Exercise 15.6

1. The following table gives the distribution of 1000 families according to the number of children. Calculate:
   (a) the arithmetic mean number of children in a family;
   (b) the standard deviation.

No. of children in family	0	1	2	3	4	5	6	7
No. of families	25	306	402	200	53	8	4	2

   (L)

2. Explain what is meant by 'dispersion' in a frequency distribution and describe clearly two methods by which it may be measured.

   (L)

3. The 'life' of an electric lamp is the number of hours it will last at its standard voltage before it breaks down. A manufacturer tests the lives of a sample 200 of the same type of lamp. The number breaking down in each successive interval of 200 hours is given in the following frequency table.

Life (in Hundreds of hours)	0–2	2–4	4–6	6–8	8–10	10–12	12–14	14–16	16–18	18–20
Number of breakdowns	1	3	8	22	46	55	38	20	5	2

   Taking a working mean of 1100 hours, or otherwise, calculate the mean life and standard deviation of this type of lamp.   (L)

4. A time study of a workman performing a certain industrial operation gave the following times in seconds. Calculate the arithmetic mean and the standard deviation correct to 2 significant figures.

   5·1, 4·9, 5·6, 6·0, 5·2, 5·8, 4·9, 5·6, 5·2, 5·7.

5. A company which manufactures tubes for television receivers conducted a test of a sample batch of 1000 tubes and recorded the number of faults in each tube in the following frequency table. Calculate the arithmetic mean and the standard deviation of the number of faults.

No. of faults	0	1	2	3	4	5	6
Frequency	620	260	88	20	8	2	2

   (L)

6. Calculate the mean and standard deviation of the following set of numbers:

5, 8, 3, 13, 2, 10, 15. (*L*)

7. The following table is an analysis of the daily wages paid by a firm:

Wage (£)	3–5	6–8	9–11	12–14	15–17	18–20	21–23	24–26
No. of employees	4	10	26	38	13	6	2	1

Calculate the arithmetic mean and the standard deviation.

8. The following numbers give the monthly amounts (in million tonnes) of coal produced in ten consecutive months. Calculate the mean and the standard deviation.

4·7, 4·6, 4·5, 4·4, 4·3, 4·2, 3·5, 3·7, 4·4, 4·7. (*L*)

9. The following table gives the distribution of marks obtained in a test by 400 candidates:

Marks	1	2	3	4	5	6	7	8	9	10
Frequency	6	24	60	86	90	68	37	22	5	2

If m_1, m_2, and σ are, respectively, the mean, the mode and the standard deviation of this distribution, find, correct to three significant figures, the value of

$$\frac{m_1 - m_2}{\sigma}$$ (*L*)

10. The following numbers give the annual totals (in hundreds) of fatal accidents in factories for the period 1948–1957. Calculate the mean and the standard deviation.

8·5, 7·6, 8·0, 8·2, 7·9, 7·4, 7·1, 7·0, 6·8, 6·5. (*L*)

11. The following numbers give the monthly amounts (in million pounds) of the repayments of accrued interest on National Savings for nine consecutive months in 1958:

6·4, 8·8, 7·1, 7·0, 6·0, 9·5, 9·5, 7·7, 6·4.

Calculate the variance of this set of numbers. (*L*)

12. The following table gives the number of motor-cyclists killed during 1957:

Age	15–20	20–25	25–30	30–40	40–50	50–60	60–70	70–80
No. killed	224	345	174	179	131	70	29	4

Working from $27\frac{1}{2}$ as an assumed mean age, calculate the mean and standard deviation of this distribution. (*L*)

13. An examination of 10 samples of a product showed the presence of the following percentages of impurity:

5·34, 4·96, 5·43, 4·87, 5·50, 5·27, 4·90, 5·04, 5·13, 5·46.

Find the mean percentage of impurity and the standard deviation of the samples. (*A.E.B.*)

14. The following figures give the monthly consumption of butter and cheese during the period January to June. Calculate (*a*) the average monthly consumption for each series, (*b*) the standard deviation of each

	Thousand Tonnes	
	Butter	Cheese
Jan.	33·3	24·8
Feb.	27·1	23·9
Mar.	27·2	21·4
Apr.	33·6	22·2
May	26·4	18·3

15. In this table, which shows the weekly salary of 120 executives in a company, '150–' means 'from 150 to 160, including 150 but excluding 160.'

Weekly salary (£)	150–	160–	170–	180–	190–	200–	210–	220–
No. of men	1	1	2	5	12	26	50	16

Weekly salary (£)	230–	240–
No. of men	5	2

Calculate
(*a*) the average salary,
(*b*) the median salary,
(*c*) the mean deviation and
(*d*) the standard deviation.

16. The table shows the intelligence quotient (I.Q.) of 100 pupils at a certain school.

Calculate
(*a*) the mean,
(*b*) the mean deviation,
(*c*) the standard deviation.

I.Q.	55–	65–	75–	85–	95–	105–	115–	125–	135–
No. of pupils	1	3	7	20	32	25	10	1	1

Note: 55– means 'from 55·0 to 64·9 inclusive', each I.Q. being given correct to one decimal place. (*CA.*)

17. The frequency distribution of the marks obtained by 400 candidates in an examination is as follows.

Marks   1–20 21–30 31–40 41–50 51–60 61–70 71–80 81–100
Frequency   9   32   89   102   78   63   21   6

(i) Calculate the arithmetic mean and the standard deviation of the distribution.

(ii) The examiner decides to publish the original marks out of a total of 500, that is each candidate's mark is multiplied by five. *Deduce* the new arithmetic mean and standard deviation. Give reasons.   (*A.E.B.*)

18. The number of gramophone records produced in each month of 1964 was:

Month	Jan.	Feb.	Mar.	Apr.	May	June
Records						
(Thousand)	10 723	8321	8077	8886	7074	7257
	July	Aug.	Sept.	Oct.	Nov.	Dec.
	7203	7034	8940	8317	9472	9953

(Monthly Digest)

Calculate

(i) the arithmetic mean number of records produced per month,

(ii) the monthly deviation from the mean,

(iii) the mean deviation from the mean.   (*A.E.B.*)

19. The numbers of questions of equal difficulty attempted by a random sample of 1645 pupils in a two-hour test are given in the table.

Number of questions   0– 3– 6– 9– 12– 15– 18– 21– 24–
Number of pupils   3 22 75 194 331 402 321 192 79
                     27– 30–
                      22   4   Total 1645

Find (i) the median, (ii) the mode, and calculate (iii) the arithmetic mean, (iv) the standard deviation.

It was decided that, in tests of equal difficulty, about $2\frac{1}{2}$ per cent of the candidates were to be given the opportunity of obtaining full marks. How many questions should the candidates be asked to answer in two hours?   (*A.E.B.*)

20. (*a*) Explain briefly the term *standard deviation*.

(*b*) (i) Calculate the mean and standard deviation for both classes of grocers in the table.

(ii) Which class of grocers varies more in size, assuming that the value of sales is a measure of size? Give statistical reasons for your answer.   (*A.E.B.*)

202

Grocers' Sales

Value of Sales	Grocers with Meat	Grocers with Off-Licence
£1000+	25	92
£2000+	63	230
£3000+	69	260
£4000+	82	249
£5000−£10000	39	116
Total	278	947

21. Calculate the arithmetic mean and the standard deviation of the distribution in the table.

Variable x 2·5–   7·5–   12·5–   17·5–   22·5   27·5–   32·5–
Frequency   14   30   67   123   177   200   178
           37·5–   42·5–   47·5–   52·5–62·5
           122   68   29   16

Draw an ogive of the distribution and use it to determine the proportion of frequencies between

(i) the arithmetic mean $\pm$ one standard deviation,
(ii) the arithmetic mean $\pm$ two standard deviations.   (A.E.B.)

22. The table gives the lifetime of 430 valves.

Lifetime (to the nearest hour)   300–399 400–499 500–599
Number of valves   20   49   61
   600–699 700–799 800–899 900–999 1000–1099 1100–1199
     83   70   64   51   23   7
   1200–1299
      2   Total 430

Determine (i) the upper limit of the fourth class, (ii) the lower limit of the seventh class, (iii) the class-interval.

Construct a smoothed ogive for the distribution and from your graph estimate (iv) the median value of the distribution, (v) the semi-interquartile range of the distribution, (vi) the number of valves expected to burn out in 550 hours, (vii) the percentage of the valves sold that the manufacturer will have to replace if he guarantees a valve to last 425 hours?   (A.E.B.)

Standardised Examination Marks

In the internal examinations of some schools, the candidates' marks in each subject are added together to give a final mark, and a final order of merit is decided from these marks. If the mean mark and the standard deviation of the marks in each subject are fairly close, the method is simple and works fairly well. But most candidates are familiar with the fact that the marks in different subjects have different arithmetic means and dispersions. Thus, candidates with high marks in subjects with large dispersions will have an advantage, and those with top marks in subjects where the dispersion is small will be at a disadvantage. Many ways of trying to eliminate this unfairness have been suggested, and the most popular way is to 'standardise' the marks in each subject. Thus, if the marks in a particular subject are x_1, x_2, ... x_n, and the mean and standard deviation of the marks are M and σ, respectively, the standardised scores are $\dfrac{x_1 - M}{\sigma,}$ $\dfrac{x_2 - M}{\sigma}$, ... $\dfrac{x_n - M}{\sigma}$, that is, the deviation of the mark from the arithmetic mean is measured in units of the standard deviation. If this is done for all subjects, the standardised scores of each candidate may then be added together and will give a better order of merit.

This method will give some negative marks and, to avoid these, it is sometimes assumed that the final mean mark for each subject is to be 50 and the final standard deviation 16. The final mark for each candidate in each subject is then

$$50 + \frac{(x - M)16}{\sigma}$$

Example

In a mathematics examination, three candidates A, B, and C obtained 37, 61 and 73 marks, respectively. The average mark for the paper was 43 and the standard deviation of the marks 12. Calculate the candidates' scores if we make a new distribution with a mean of 50 and a standard deviation of 16.

Candidate:	A	B	C
Standardised scores:	$\dfrac{37 - 43}{12} = -0.5$	$\dfrac{61 - 43}{12} = 1.5$	$\dfrac{73 - 43}{12} = 2.5$
Final scores:	$50 - 16 \times 0.5 = 42$	$50 + 16 \times 1.5 = 74$	$50 + 16 \times 2.5 = 9$

Example

Candidate A received a mark of 87 in a mathematics examination where the arithmetic mean and standard deviation of all the marks were 60 and 9, respectively. The same group of candidates sat a physics examination, and candidate A received a mark of 93. The mean and standard distribution of the physics marks were 84 and 15, respectively. In which examination does candidate A show the better performance?

Standardised scores: Mathematics $= \dfrac{87 - 60}{9} = 3$; Physics $= \dfrac{93 - 84}{15} = 0.6$.

The better performance is in mathematics.

Linear Transformations of the Mean and Standard Deviation

The formula for standardising examination marks given on page 204 is derived from the properties of the mean and standard deviation of a distribution and how they are affected when all the individual values of the distribution have a constant amount added to or subtracted from them, or multiplied or divided by a constant. The mathematical principles involved are:

1. The addition or subtraction of a constant from each of the values of the distribution alters the mean of the distribution by the same amount but the standard deviation is not affected.

2. If each value of the distribution is multiplied or divided by the same amount then **both** the mean and the standard deviation are changed in the same manner.

Thus, if we wish to transform the mean and the standard deviation we must first operate on the standard deviation by multiplying all the data by the ratio of the required s.d. to the original s.d. Thus if

the required s.d. $= \sigma_r$ the original s.d. $= \sigma_o$

the required mean $= M_r$ the original mean $= M_o$

and each of the original values of the distribution is represented by x_i, then after the first operation:

the σ_o is transformed to σ_r,

all x_i become $\dfrac{\sigma_r}{\sigma_o} x_i$,

and at this stage M_o becomes $M_o \dfrac{\sigma_r}{\sigma_o}$.

Hence to obtain the required mean we must add to all the newly transformed data the difference between the required mean and the newly transformed original mean;

i.e. we must add $M_r - M_o \dfrac{\sigma_r}{\sigma_o}$ to each $x_i \dfrac{\sigma_r}{\sigma_o}$ which gives

$$x_i \frac{\sigma_r}{\sigma_o} + \left(M_r - M_o \frac{\sigma_r}{\sigma_o} \right) = M_r + x_i \frac{\sigma_r}{\sigma_o} - M_o \frac{\sigma_r}{\sigma_o}$$

$$= M_r + \frac{(x_i - M_o)\sigma_r}{\sigma_o}$$

which the reader will recognise as the formular given on p. 204.

Example

An examination in French consists of two parts, oral and written, which were regarded as being of equal importance. An analysis of the mark distribution gave the following result.

	Arithmetic Mean	Standard Deviation
Oral	42	16
Written	53	32

Subsequently, each set of marks was adjusted to an arbitrary scale with arithmetic mean 50 and standard deviation 20, before adding together to produce a set of final marks for the subject.

Calculate

  (i) the arithmetic mean of the set of final marks,

 (ii) the final mark of a candidate who scored 50 in the oral section and 37 in the written section,

(iii) the final mark of a candidate who scored 34 in the oral section and 69 in the written section.

State briefly the benefit to be gained by such a procedure.   (*A.E.B.*)

  (i) Since each section is to have a final mean of 50 and the sectional marks are added, the mean of the combined sections is $50 + 50 = 100$.

 (ii) this could be answered by substitution in the formula, but it will be solved here step by step to illustrate the work shown in the derivation of the formula.

Ratio of s.d.'s $= \dfrac{\sigma_r}{\sigma_o} = \dfrac{20}{16} = \dfrac{5}{4}$ $\therefore$ all oral marks must first be multiplied by this factor.

This will make the original mean become $42 \times \dfrac{5}{4} = 52 \cdot 5$

Hence we must add $50 - 52 \cdot 5 = -2 \cdot 5$ to each value to get a mean of 50.

Similarly for the written marks;

s.d. ratio $= \dfrac{20}{32} = \dfrac{5}{8}$,

the mean becomes $53 \times \dfrac{5}{8} = \dfrac{265}{8} = 33\dfrac{1}{8}$,

hence we must add $50 - 33\dfrac{1}{8} = 16\dfrac{7}{8}$ to each value.

50 in oral becomes $50 \times \dfrac{5}{4} - 2 \cdot 5 = 60$

37 in written becomes $37 \times \dfrac{5}{8} + 16\dfrac{7}{8} = 40$

$$\text{Total} = 60 + 40 = 100$$

(iii) 34 in oral becomes $34 \times \frac{5}{4} - 2 \cdot 5 = 40$

69 in written becomes $69 \times \frac{5}{8} + 16\frac{7}{8} = 60$

Total $= 60 + 40 = 100$

Exercise 15.7

1. n candidates sit for an examination. The arithmetic mean of their marks is M and the standard deviation σ. The marks are converted to a new scale by the formula

$$y = 50 - \frac{20(M - x)}{\sigma}$$

Find the mean and standard deviation of the new marks.

2. If the 'crude' mark x of each of the n pupils in a class is replaced by a 'standardised' mark $(x - m)/\sigma$, where m is the mean mark and σ the standard deviation of the marks in that particular subject, calculate (i) the mean mark, (ii) the standard deviation of the standardised marks.

   By standardising the crude marks given in the following table, place the four boys A, B, C, D in an order of merit for tests in the three subjects English, Mathematics, and French, taken in the same examination.

	m	σ	A	B	C	D
English	55	10	60	50	60	70
Mathematics	70	4	80	60	70	80
French	60	12	50	80	60	40	(C.A.)

3. In a certain examination the maximum mark obtainable was 80 and the marks of three candidates A, B and C were 32, 48 and 76 respectively. The arithmetic mean of all the marks was 45 and their standard deviation was 20. If the marks are expressed as percentages, what do the arithmetic mean and the standard deviation become?

   If the marks are adjusted, without their relative value being altered, so that the arithmetic mean becomes 50 and the standard deviation 24, find the adjusted marks of A, B and C. (L)

4. The marks obtained by a particular group of five pupils in a class of forty were as follows:

Pupil	A	B	C	D	E
Mathematics	56	76	58	44	57
English	71	59	63	87	71

207

The mean and standard deviation for all the pupils in Mathematics and English were

	Mean for all pupils	Standard deviation for all pupils
Mathematics	42	15
English	57	12

Some adjustment of the marks was made by the examiner.

 (i) He decided to make the overall arithmetic mean mark for each set of 40 papers equal to 50 by adding a constant to each mark. Calculate the adjusted marks for the 5 pupils.

(ii) He then decided to scale the marks obtained in (i) so that the standard deviation for each paper was 15.

Using the marks standardised as in (i) and (ii) above, calculate the average mark for each of these 5 pupils and say who you would place at the top among these five.

Given that the Mathematics examination lasts 2 hours and the English examination 1 hour, recalculate the average mark for each pupil, weighting the standardised marks according to the duration of the examination. Does the same pupil still come out top?

If pupil E had had an average mark of 72, what would have been the relative weights of the standardised marks in Mathematics and English?

Further Examples

These examples may be left for a second reading. They show the ease with which the arithmetic mean and standard deviation can be manipulated, and illustrate why they should be used in preference to other methods of measuring means and dispersion.

Example

The arithmetic mean and the variance of three numbers x, y, and z is 5 and $\frac{8}{3}$, respectively, and the arithmetic mean and variance of the six numbers x, y, z, a, b, and c is 8 and $\frac{35}{3}$, respectively. What is the arithmetic mean and variance of the numbers a, b, and c?

The mean of the three numbers x, y, and z is 5.

Therefore $x+y+z = 15$.

The mean of the six numbers x, y, z, a, b, and c is 8.

Therefore $x+y+z+a+b+c = 48$.

Therefore $a+b+c = 48-15 = 33$.

Therefore the mean a, b, and c is $\frac{33}{3} = 11$.

To find the variance, use the formula on page 194, that is, the variance of a distribution equals the arithmetic mean of the squares of the values of the variable minus the square of the arithmetic mean of the values of the variable.

Then, $\qquad \dfrac{8}{3} = \dfrac{x^2 + y^2 + z^2}{3} - 25$

Therefore, $\quad x^2 + y^2 + z^2 = 83$

Also, $\qquad \dfrac{35}{3} = \dfrac{x^2 + y^2 + z^2 + a^2 + b^2 + c^2}{6} - 64$

Therefore, $\quad x^2 + y^2 + z^2 + a^2 + b^2 + c^2 = 454$

and $\qquad\qquad\qquad\quad a^2 + b^2 + c^2 = 371$

Therefore the variance of a, b, and c

$$= \dfrac{371}{3} - 121 = \dfrac{8}{3}$$

Example

The numbers of members, means and standard deviations of three distributions are

Number of members	280	350	630
Means	45	54	49
Standard deviations	6	4	8

Find the mean and standard deviation of the distribution formed by the three distributions taken together.

The total number of items in the combined distribution is
$$280 + 350 + 630 = 1260$$
The sum of the items in the combined distribution is
$$280 \times 45 + 350 \times 54 + 630 \times 49 = 62370$$

Therefore the mean of the combined distribution is $\dfrac{62370}{1260} = 49 \cdot 5$.

Using the formula on page 194:

Distribution 1.

$$36 = \frac{\text{Sum of the squares of the values of the variable}}{280} - 2025$$

Therefore the sum of the squares of the values of the variable
$$= 577080$$

Distribution 2.

$$16 = \frac{\text{Sum of the squares of the values of the variable}}{350} - 2916$$

Therefore the sum of the squares of the values of the variable
$$= 1026200$$

Distribution 3.

$$64 = \frac{\text{Sum of the squares of the values of the variable}}{630} - 2401$$

Therefore the sum of the squares of the values of the variable
$$= 1\,552\,950$$

Therefore, the sum of the squares of the values of the variable in the combined distribution $= 577\,080 + 1\,026\,200 + 1\,552\,950 = 3\,156\,230.$

Therefore, $\sigma^2 = \dfrac{3\,156\,230}{1260} - (49\cdot5)^2 = 54\cdot69$

Therefore, $\sigma = 7\cdot40.$

(The last example is from an advanced level Applied Mathematics paper of London University. The examples marked (*L.A.*) in the following exercise are also from advanced level papers, but their solution should not be difficult for anyone who followed the solutions of the two examples.)

Exercise 15.8

1. The arithmetic mean of n numbers is M. The sum of the first $(n-4)$ numbers is S. Calculate the arithmetic mean of the last four numbers.

2. If $z_1 = ax_1 + by_1,$
    $z_2 = ax_2 + by_2,$
    . . . .
    . . . .
    $z_n = ax_n + by_n.$
    show that $M = aM_1 + bM_2$
    where M is the arithmetic mean of $z_1, z_2, \ldots z_n,$
    $\quad\quad M_1$,, ,,   ,,   ,,   ,, $x_1, x_2, \ldots x_n,$
    and M_2 ,, ,,   ,,   ,,   ,, $y_1, y_2, \ldots y_n.$

3. The mean of n numbers $x_1, x_2, \ldots x_n$ is M. A new series of numbers $y_1, y_2, \ldots y_n$ is formed where
    $$y_1 = (a+x_1)^2 - x_1^2,$$
    $$y_2 = (a+x_2)^2 - x_2^2,$$
    . . . .
    . . . .
    $$y_n = (a+x_n)^2 - x_n^2.$$
    Show that the mean of $y_1, y_2, \ldots y_n$ is
    $$a(a+2M)$$

4. The mean of n_1 numbers is M_1, and the mean of $(n_1 + n_2)$ numbers is M. Show that the mean of the n_2 numbers is

$$M + \frac{n_1}{n_2}(M - M_1)$$

5. If s and σ are the root-mean-square deviation and the standard deviation, respectively, of a set of numbers x_1, x_2, ...x_n, write down expressions for s^2 and σ^2, given M and W are the arithmetic mean and working mean of the numbers, respectively.

By substitution show that

$$s^2 - \sigma^2 = (M - W)^2$$

6. The numbers of members, means and standard deviations of two distributions are

Number of members	100	200
Means	42	45
Standard deviations	5	3

Find the mean and standard deviation of the distribution formed by the two distributions taken together.

7. The first term of an arithmetic progression is a and the common difference is d. Calculate the mean and variance of the first three terms. Repeat for the first 5, 7, and 9 terms. From a consideration of the variance of these four sets of numbers, *deduce* the variance for the first 11 and 13 terms.

8. If M and σ are the mean and standard deviation of a distribution, and s is the root-mean-square deviation from a working mean of zero, show that

$$\sigma^2 = s^2 - M^2$$

Using this relation calculate the standard deviation of the following set of numbers:

6·3, 7·4, 5·3, 1·5, 3·2, 4·6, 2·9, 8·3, 9·7, 1·8.

9. Using the relation derived in Question 8, find the mean and standard deviation of the numbers

(a) 1, 2, 3, ...n,

(b) 1^2, 2^2, 3^2, ...n^2

given,

$$1^2 + 2^2 + \ldots + n^2 = \frac{1}{6}n(n+1)(2n+1)$$

$$1^4 + 2^4 + \ldots + n^4 = \frac{1}{30}n(n+1)(6n^3 + 9n^2 + n - 1).$$

10. Find the arithmetic mean of

$$1, 2, 4, 8, \ldots 2^{n-1}.$$

11. Using the relation
$$\sigma^2 = s^2 - M^2$$
calculate the variance of
$$1, 2, 4, 8, \ldots 2^{n-1}.$$

12. Find the mean and variance of
$$a, ar, ar^2, ar^3, \ldots ar^{n-1}.$$

13. The mean of the six numbers 6, 9, 3, 2, x, y is 6 and the variance is 10. Find the values of x and y. (L)

14. The mean of two numbers x and y is 5 and the standard deviation is 2. The mean of five numbers x, y, a, b, and c is 8 and the standard deviation is 4. Find the mean and the standard deviation of the three numbers a, b, and c. (L)

15. Three numbers X_1, X_2, and X_3 have a mean $\bar{X}$. If x_1, x_2, and x_3 are, respectively, the deviations of X_1, X_2 and X_3 from $\bar{X}$, prove that
$$x_1{}^2 + x_2{}^2 + x_3{}^2 = X_1{}^2 + X_2{}^2 + X_3{}^2 - 3\bar{X}^2.$$

Write down the corresponding result for n numbers X_1, X_2, $X_3, \ldots X_n$. (L)

16. Find the arithmetic mean of each of the following distributions:

| Intelligence Test | No. of Candidates | |
Score	Boys	Girls
30–	3	⎱ 6
35–	17	⎰
40–	36	21
45–	52	43
50–	76	60
55–	61	34
60–	43	⎱ 12
65–70	18	⎰

Verify that the mean for boys can also be found by the following method: in a separate column cumulate the frequencies from the bottom upwards, stopping one row from the top; let S stand for the sum of this column. The mean is then given by
$$x_1 + \frac{hS}{n},$$
where h is the class-interval and x_1 is the mid-point of the first class. Suggest reasons why this method fails for the distribution of girls. (L.G.I.)

212

16 Scatter Diagrams and the Equation of a Straight Line

Degrees of Relationship

Consider the following examples:

Example 1

The values of y corresponding to the values of x in the equation $y = 2x + 1$, are given in the following table:

x	0	1	2	3	4
y	1	3	5	7	9

and if these are plotted on a graph they lie on a straight line, because there is a definite linear relationship between x and y, namely

$$y = 2x + 1.$$

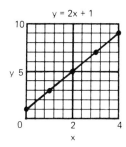

Example 2

In an experiment on pulleys to find what force is required to lift a given weight, the following results were obtained:

W = weight in newtons	10	20	30	40	50
P = force in newtons	3·2	4·8	7·2	9·2	10·8

When these points are plotted on a graph we see that they lie approximately on a straight line, and we draw the line of 'best fit' and assume there is an approximate relation between P and W similar to the relationship between x and y of the last example; i.e. a relationship of the form $P = aW + b$, where a and b are constants. Generally in mechanical experiments the points lie very near the straight line, and values of P read off from the graph agree very closely with the results obtained by experiment.

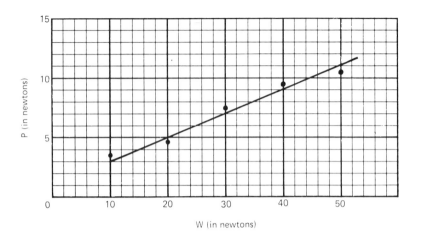

Example 3

Below are given the marks out of a total of 50 in French and German awarded in two tests to a group of boys taught by the same master:

| French | 10 | 10 | 18 | 25 | 28 | 33 | 34 | 39 | 42 | 43 |
| German | 11 | 22 | 22 | 19 | 35 | 27 | 33 | 40 | 42 | 47 |

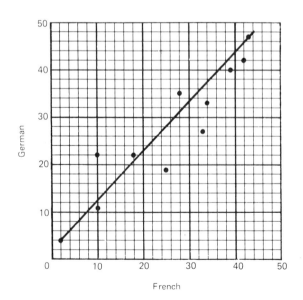

214

If these marks are plotted on a graph, say the French marks along the horizontal axis and the corresponding German marks along the vertical axis, there is a greater scatter of points than in the previous example. But though the points are more widely scattered it can be concluded that generally high and low marks in French are associated with high and low marks, respectively, in German. We assume that there is some linear association between the two sets of marks, and we draw the line of best fit. Although we draw the general conclusion stated above, we cannot make the statement that any particular boy will receive high or low marks in both languages. The relationship is not so close as in the mechanical experiment.

Example 4

Tests were given to a group of boys in Chemistry and English, and the marks obtained out of 50 by each boy are tabulated below:

Chemistry   4 11 13 17 21 22 31 32 33 37 40 45 45 47 49 49
English   40 10 26 46 20   9 40 22 14 44 10 30 40 10 28 44

Now, when these points are plotted on a graph, there is a large degree of scatter and no pattern emerges. We conclude that there is no relationship between the marks obtained by this group of boys in English and Chemistry.

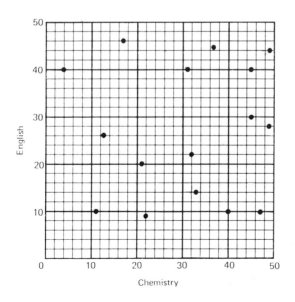

215

Our examples have shown us degrees of relationship between two variables, varying from the exact, through the not so exact, to no relationship at all. When the points on the graph form a pattern which approximates to a narrow straight belt, the variables are said to be *linearly correlated.* If large and small values of the first variable correspond to large and small values of the second variable, respectively, the variables are said to be *directly correlated.* If large and small values of the first variable correspond to small and large values of the second variable, respectively, the variables are said to be *inversely correlated.* The graph is called a *scatter diagram,* and the line of best fit the *regression line.* The regression line shows how the subject varies with the relative, and the slope of the regression line is called the *regression coefficient.*

(The slope of the line is the tangent of the angle the regression line makes with the positive direction of the axis of the relative. That is, if a right-angled triangle ABC is drawn, with right-angle at B, the regression coefficient is $\dfrac{AB}{BC}$ where AB and BC are measured on the scales of the items on the graph.)

Little or No Scatter

In Example 2 the points all lie in a very narrow band, which shows a high degree of correlation, and it is easy to draw the line of best fit. This line can be used to find values of P for given values of W, and values of W for given values of P. Thus, if $W = 25$ we find from the graph $P = 6 \cdot 1$, and if $P = 8$, $W = 34 \cdot 5$.

Example

The following are observed values of two quantities x and y. Plot these points and, if there is a high degree of correlation, draw the line of best fit.

| x | 15·0 | 22·6 | 30·0 | 37·0 | 45·0 | 50·0 | 60·0 |
| y | 3·76 | 6·60 | 9·60 | 12·2 | 15·0 | 17·2 | 21·0 |

Find (i) the regression coefficient

       (ii) the value of y when $x = 21$

      (iii) the value of x when $y = 17 \cdot 5$.

The points lie in a very narrow band and therefore there is a high degree of correlation.

(i) The regression coefficient $= \dfrac{AB}{BC} = \dfrac{21 - 4}{60 - 15} = \dfrac{17}{45} = 0 \cdot 38.$

(ii) The value of y when $x = 21$ is 6.

(iii) The value of x when $y = 17 \cdot 5$ is 51.

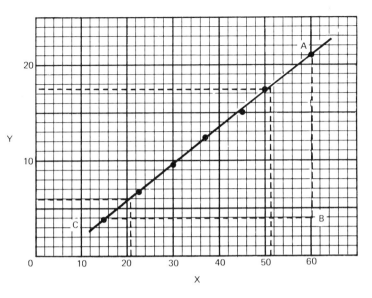

Exercise 16.1

If the following sets of corresponding values of x and y show a high degree of correlation, draw the line of best fit, and find

    (a) the regression coefficient.

    (b) the value of y when $x = 13$.

    (c) the value of x when $y = 25$.

1.	x	2	4	6	8	10	12	14	16	18
	y	20	30	40	50	60	70	80	90	100

2.	x	10	30	40	60	70	80	90	100
	y	37	49	55	67	73	79	85	91

3.	x	4	8	10	12	16	20	24	28	30
	y	53	45	41	37	29	21	13	5	1

4.	x	23	31	40	49	59	72
	y	14	20	26	35	44	54

5.	x	4	10	16	22	28	34	40
	y	21	31	41	51	61	71	81

Fair Degree of Scatter

If there is a fair degree of scatter of the points, there are two regression lines. Some students find it difficult to understand the necessity for two regression lines, but the following explanation may help to make it clear.

Suppose the following table gives the marks obtained by 15 boys on their English and French examination papers:

| English | 12 | 24 | 30 | 34 | 47 | 58 | 68 | 50 | 50 | 50 | 30 | 36 | 38 | 58 | 30 |
| French | 32 | 44 | 47 | 58 | 73 | 72 | 88 | 60 | 50 | 80 | 30 | 44 | 70 | 78 | 61 |

and that a sixteenth boy obtains 44 marks in the French examination and is absent for the English examination. A rough estimate of his English mark can be obtained by finding the arithmetic mean of the English marks of the boys who obtained 44 marks in the French examination. These marks are 24 and 36 and their arithmetic mean is 30.

If another boy is absent for the examination in French, but obtains 30 marks for the examination in English, a rough estimate of his mark in French can be obtained by finding the arithmetic mean of the French marks of the boys who obtain 30 marks in the examination in English. These are 47, 30, and 61, and their arithmetic mean is 46.

If there were sufficient examinees, we could do this for each mark in French and each mark in English, and would obtain two sets of averages, one to give us the English mark corresponding to a given French mark, and the other to obtain the French mark corresponding to a given English mark. Thus we see we need two regression lines. In the particular example above we find the average of two or three particular points, but the regression lines give us the best average using all the points.

Aid to Drawing the Regression Line
When the points are plotted on a graph, it is difficult to judge the position of the regression lines with accuracy, and it is helpful to obtain some guidance. It can be proved that the regression line must pass through the mean of the marks. Therefore, first calculate this mean.

Mean of the English marks $= \dfrac{615}{15} = 41$

Mean of the French marks $= \dfrac{887}{15} = 59 \cdot 1$

Plot this point (M).

Imagine a line drawn through this point parallel to the axis showing the English marks. This divides the set of points into two arrays and the regression line should pass near the means of these two arrays. The points to the left of this line are:

| English | 30 | 12 | 36 | 24 | 30 | 50 | 34 |
| French | 30 | 32 | 44 | 44 | 47 | 50 | 58 |

and to their means are 30·9 and 43·6, respectively.

To the right of the line the points are:

| English | 50 | 30 | 38 | 58 | 47 | 58 | 50 | 68 |

218

French   60 61 70 72 73 78 80 88

and their means are 49·9 and 72·8, respectively.

Plot these two points (*P* and *Q*).

The line drawn through *M* and as near as possible through the points *P* and *Q* is the line of regression of the English marks on the French marks, and is used to find the English mark corresponding to a given French mark.

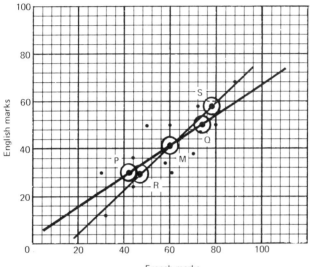

French marks

Thus a boy receiving 54 marks in French receives, on the average, 38 marks in English.

To find the regression line of the French marks on the English marks imagine a line through the point (*M*) parallel to the axis showing the French marks, and dividing the set of points into two arrays. Find the means of the points above and below this line. The points below the line are:

English   12 30 36 24 30 34 30 38

French   32 30 44 44 47 58 61 70

and their means are 29·3 and 48·3, respectively.

The points above the line are:

English   50 50 58 47 58 50 68

French   50 60 72 73 78 80 88

and their means are 54·4 and 71·6, respectively.

Plot these points (*R* and *S*), draw a line through *M* and as near as possible through *R* and *S*. This line is used to find the French mark corresponding to a given English mark. Thus 69 is the average mark in French corresponding to a mark of 50 in English.

When there is a degree of scatter, there are always two regression lines, and it is essential to be clear which one is to be used in a particular problem.

Example

The following are the marks awarded to 20 pupils in examinations in Woodwork and Music. Draw the regression lines.

Music	5	7	11	20	23	25	30	31	31	35
Woodwork	23	93	45	80	65	11	90	9	27	55

Music	45	50	60	67	79	79	85	93	93	93
Woodwork	85	40	70	15	47	83	65	13	39	83

The mean of the marks is Music 48·1 and Woodwork 51·9.
Plot this point (M).

The mean of the points to the left of M is Music 23·9, and Woodwork 53, and to the right, Music 77·7 and Woodwork 50·6.

Plot these two points (P and Q).

The line through M and as near as possible through P and Q is the line of regression of the Woodwork marks on the Music marks.

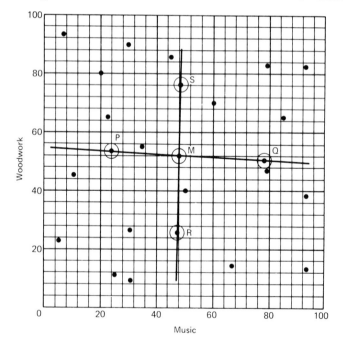

The mean of the points below M is Music 48·5, and Woodwork 26·9, and the mean of the points above M is Music 47·7, and Woodwork 76·9.

Plot these two points (R and S).

The line through M and as near as possible through R and S is the line of regression of the Music marks on the Woodwork marks.

One glance at the scatter diagram tells us there is no correlation between the Music marks and the Woodwork marks, and in this case the regression lines are nearly at right angles.

The smaller the angle between the regression lines, the closer the correlation.

To allow for simplicity in explanation, the number of items in the distributions has been kept small, and it would be wrong to draw any general conclusions from such small distributions.

Exercise 16.2

The following are the marks obtained by candidates in an examination in two subjects. In each case plot a scatter diagram and draw

(a) the regression line of subject **A** on subject **B**,
(b) the regression line of subject **B** on subject **A**.

1. Subject **A**   7 20 30 53 64 66 75 81 94
   Subject **B**   9 30 21 30 50 60 63 80 83

2. Subject **A**   1   7 10 15 17 24 38 42 45
   Subject **B**   3   5 14 11 17 19 25 36 39

3. Subject **A**   8   9 10 13 17 20 21
   Subject **B**   12   4 14 10   8 20 19

4. Subject **A**   10 29 43 56 64 75 83 90 42 51 63 74 85 93
   Subject **B**   21   7 55 48 70 41 68 91 43 55 31 68 87 90

5. Subject **A**   4   7   9 10 15 20 24 26 30 37 41 45 49 50 51 53 55 60
   Subject **B**   8   6 11   7 20 40 30 30 21 42 45 38 57 46 59 50 63 54

6. The following are the marks of 10 candidates in two subjects in an examination:
   Subject **A**   35 53 47 38 56 43 64 73 62 47
   Subject **B**   39 51 50 39 51 46 67 71 69 52

   Plot a scatter diagram and draw the regression lines.

   A candidate obtained 60 marks in subject **A** but was absent for the examination in subject **B**. Estimate his mark in subject **B**.

   Another candidate obtained 35 marks in subject **B** but was absent for the examination in subject **A**. Estimate his mark in subject **A**.

7. The average daily cost, C pounds, of running a bus when N passengers are carried is given in the following table:

C	56	67	72	76	77	80	86	91
N	258	525	620	735	755	783	920	1020

Plot the points and see if there is a linear relationship between C and N.

Variables Associated More than Once

When the variables are associated in pairs, as in the examples, the distribution is known as a bivariate distribution. If the number of associations is large there will be some pairs of variables associated more than once. It is difficult to plot several coincident points on a graph, and the scatter diagram breaks down. We then proceed as in the following examples:

Example

The following table gives the results of two tests given to 114 pupils in Physics and Mathematics arranged as a bivariate distribution.

		Physics and Mathematics Marks										Mean of Row
	10	–	–	–	–	–	–	–	–	–	–	0
	9	–	–	–	–	–	–	–	–	1	1	9·5
	8	–	–	–	–	–	2	4	3	1	–	7·3
	7	–	–	–	1	3	3	3	4	1	–	6·6
Physics	6	–	–	1	3	6	5	4	2	–	–	5·7
Marks	5	–	–	1	6	6	4	1	–	–	–	4·9
	4	1	2	4	6	4	2	1	–	–	–	4·0
	3	1	3	4	2	2	2	–	–	–	–	3·5
	2	1	2	2	3	2	1	–	–	–	–	3·5
	1	1	1	1	–	–	–	–	–	–	–	2·0
		1	2	3	4	5	6	7	8	9	10	

Mathematics Marks

Mean of Col. 2·5 2·75 3·4 4·3 4·9 5·4 6·6 7·1 8·0 9·0
Mean Physics Mark = 5   Mean Mathematics Mark = 5

The table tells us that of the pupils who obtained 5 marks in Mathematics, three obtained 7 marks, six 6 marks, six 5 marks, four 4 marks, two 3 marks, and two 2 marks in Physics, and so on. The mean Physics mark obtained by pupils awarded 5 marks in Mathematics is obtained from this column, and is 4·9. If we find the means of the Physics marks in each column and plot these means against the corresponding Mathematics mark, and then draw the line of best fit through the points, we can use this line to find the probable Physics mark corresponding to a given Mathematics mark. This is the line of regression of the Physics mark on the Mathematics mark.

If we require to find the most probable Mathematics mark corresponding to a given Physics mark, we find the means of the Mathematics mark in each row, and plot these means against the corresponding Physics mark. The line of best fit through these points is the line of regression of the Mathematics mark on the Physics marks.

(*Note:* A univariate distribution is represented diagrammatically by a histogram. To represent a bivariate distribution in an analogous manner requires three axes—two horizontal axes to measure the Physics and Mathematics marks, and a vertical axis to measure the frequency.)

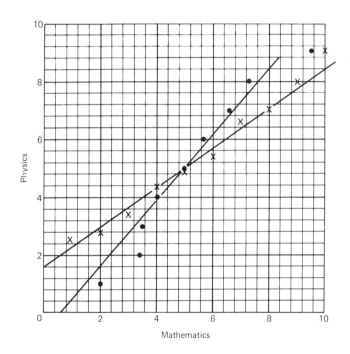

Example

The following table gives the results of two tests given to 160 pupils in English and Woodwork arranged as a bivariate distribution. Proceeding in the same way with this distribution as in the last example, we obtain the graph shown.

Woodwork and English Marks

Woodwork Mark	English 1	2	3	4	5	6	7	8	9	10	Mean of Row
10	–	–	–	–		–	–	–	–	–	0
9	–	–	3	–	1	1	–	2	1	1	6·1
8	–	–	–	2	–	1	2	–	–	–	5·6
7	3	3	–	–	11	–	3	2	1	–	4·8
6	–	–	2	13	–	11	–	2	2	1	5·4
5	2	1	–	–	12	13	3	1	2	–	5·6
4	–	–	–	11	–	12	–	1	1	–	5·3
3	–	2	2	–	12	–	2	1	2	–	5·2
2	1	1	–	3	–	1	–	1	1	–	4·8
1	–	–	1	–	1	–	2	–	–	–	5·5
Mean of Col.	5·5	4·9	5·8	5·0	4·9	5·1	5·0	5·8	5·0	7·5	

English Mark

Mean Woodwork Mark = 5·1 Mean English Mark = 5·3

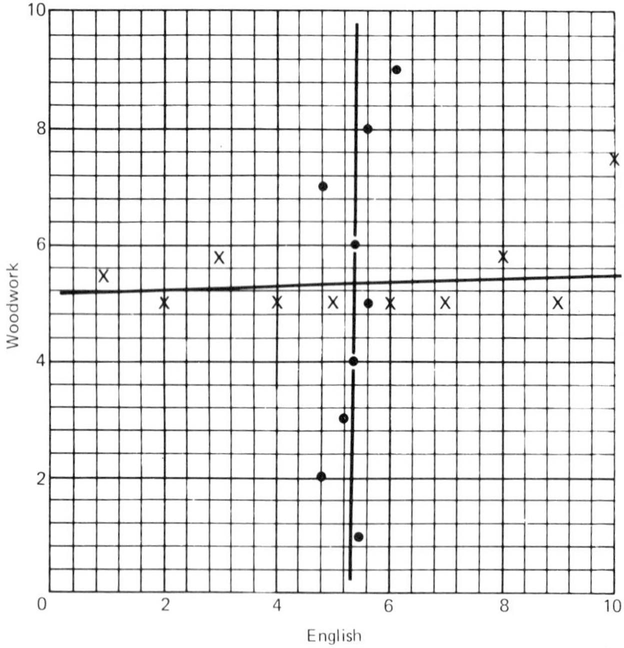

It is obvious from the graphs that, while there is correlation between the Physics and Maths marks, there is none between the English and Woodwork marks.

Exercise 16.3

Draw the regression lines for the following bivariate distributions:

1. y

5	1	1	1	1	1
4	1	1	1	1	1
3	1	1	1	1	1
2	1	1	1	1	1
1	1	1	1	1	1
	1	2	3	4	5	x

2. y

5	–	–	–	–	1
4	–	–	–	1	–
3	–	–	1	–	–
2	–	1	–	–	–
1	1	–	–	–	–
	1	2	3	4	5	x

3. y

8	–	–	–	–	–	–	–	1
7	–	–	–	–	–	–	1	2
6	–	–	–	–	1	2	1	–
5	–	–	–	–	2	3	2	–
4	–	–	1	2	2	1	–	–
3	–	1	2	2	1	–	–	–
2	1	2	1	1	–	–	–	–
1	2	1	–	–	–	–	–	–
	1	2	3	4	5	6	7	8	x

4. y

10	–	–	–	–	–	–	–	–	1	1
9	–	–	–	–	–	–	–	2	2	–
8	–	–	–	–	–	–	2	1	1	1
7	–	–	–	–	1	–	2	1	–	–
6	–	–	–	1	2	2	1	1	–	–
5	–	–	1	2	3	2	–	–	–	–
4	–	–	1	2	2	1	–	–	–	–
3	–	1	2	2	–	–	–	–	–	–
2	1	1	1	–	–	–	–	–	–	–
1	–	1	–	–	–	–	–	–	–	–
	1	2	3	4	5	6	7	8	9	10	x

5. y

y					
5	1	1	1	1	4
4	1	1	1	4	1
3	1	1	3	1	1
2	1	3	1	1	1
1	2	1	1	1	1
	1	2	3	4	5 x

6. From the school records obtain the heights, weights, and ages of each pupil in the form. From the results tabulate three bivariate distributions, i.e. height and weight, height and age, weight and age. What do you conclude?

7. The following table shows the examination marks of seven students in French and German. Construct a scatter diagram and draw a line of best fit. State any conclusion that might be drawn. Could any *general* conclusion be reached? Give a reason.

French	12	24	30	34	47	58	68
German	32	44	47	58	73	72	88

(L)

8. The following table gives the marks of eight students in each of two examinations in Botany and Zoology. Construct a scatter diagram and draw a line of best fit. State, with a reason, any (a) particular, (b) general conclusion you could reach.

Botany	39	61	49	64	42	72	52	57
Zoology	44	62	54	70	46	76	60	64

(L)

9. The following table shows the marks of 10 candidates in examinations in March and June:

Candidate	A	B	C	D	E	F	G	H	I	J
March mark	32	51	47	35	55	41	63	72	60	45
June mark	37	53	50	38	50	45	64	70	70	50

Plot a scatter diagram and draw a line of best fit.

Another candidate from this group obtained 60 marks in March, but did not sit the June examination. Find an estimated mark for him in June. (L)

10. The percentage marks for ten candidates for Intelligence and Arithmetic tests were as follows:

Candidate	A	B	C	D	E	F	G	H	I	J
Intelligence test	27	39	48	42	47	55	35	30	32	45
Arithmetic test	40	44	65	67	80	84	60	57	46	55

Plot a scatter diagram and draw the line of best fit.

A candidate **X** scored 40 in the Intelligence test and a candidate **Y** scored 55 in the Arithmetic test. Each was absent from the other test. Use your diagram to estimate a comparison between the performances of **X** and **Y**. (*L*)

1. The following data were obtained in a three-month survey of eighteen cotton mills. For each mill is given the log count (a numerical characteristic of the type of yarn spun at the mill), together with the observed mean production time per unit length.

Log count	Hours per unit length	Log count	Hours per unit length
1·3	1·6	1·4	1·3
1·5	1·1	1·4	1·5
1·5	1·4	1·3	1·7
1·8	0·8	1·4	1·6
1·7	0·9	1·6	1·0
1·2	1·8	1·5	1·3
1·2	1·6	1·6	1·2
1·4	1·5	1·6	1·3
1·7	1·1	1·8	0·9

Plot a scatter diagram for the data, and sketch in by eye the line of regression of mean production time per unit length on log count. Hence determine (*a*) its approximate slope, (*b*) its ordinate for a log count of 1·5.

Describe briefly in everyday language exactly what the regression line tells you about the relationship between the two variables.

(*J.M.B.*)

2. Tarsal–metatarsal distances for 5-week-old chicks reared on vitamin-supplemented diets:

Daily dose	Diet vitamin supplement									
	Standard					Trial				
1·2	137,	143,	142,	139,	141	136,	135,	130,	132,	132
2·0	132,	124,	127,	127,	126	128,	128,	126,	126,	123
3·3	106,	110,	107,	110,	110	111,	111,	114,	113,	109

Draw separate diagrams to illustrate the effect of the daily dose on tarsal-metatarsal distance for chicks receiving the two kinds of vitamin supplement. Sketch the appropriate regression line on each diagram.

With the aid of the regression lines, deduce (i) the daily dose at which the two kinds of supplement are equivalent, (ii) the dose of the trial supplement which is equivalent to a dose of 2·0 units per day of the standard. (*J.M.B.*)

227

13. The following are the marks of 10 candidates in two subjects in an examination:

Subject A	5	10	11	20	24	25	28	32	40	45
Subject B	13	8	18	25	22	25	25	24	35	35

Plot a scatter diagram and draw the line of best fit.

A candidate obtained 27 marks in subject **A** but was absent for the examination in subject **B**. Estimate his mark in subject **B**.

(*A.E.B.*)

14. A general intelligence test was given to fifteen boys of different ages and the results are shown in the table below.

Boy	A	B	C	D	E	F	G	H	I	J
Age (months)	134	141	154	144	163	164	175	185	175	188
Marks	62	63	64	62	72	69	80	76	72	82

	K	L	M	N	O
	185	202	205	202	200
	80	82	82	85	81

Plot a scatter diagram of the number of marks against the age in months.

Draw a line of regression of marks on age.

State, with reasons, which boy should be considered the most intelligent, and which the least intelligent, taking age into consideration.

(*A.E.B.*)

15. Sixteen boys were each given 4 problems in arithmetic and 4 problems in algebra. For each problem correctly answered they were given one mark, with the following result:

Candidate	A	B	C	D	E	F	G	H	I	J	K	L	M	N	O	P
Arithmetic	3	3	2	4	1	3	3	2	2	2	4	3	3	1	2	2
Algebra	3	2	2	3	2	3	4	2	1	3	4	2	3	1	3	2

Group the candidates by their arithmetic mark and find the average algebra mark for each group.

Draw the regression line of the algebra mark on the arithmetic mark.

(*L.*)

16. Twenty boys took two papers in a Mathematics examination and their percentage marks were as follows:

Paper I	51	42	42	54	39	27	72	57	42	42	36	33	42	75	39
Paper II	60	51	45	60	30	30	57	57	39	54	39	21	36	72	45

	57	78	42	42	60
	54	78	42	48	63

228

Plot a scatter diagram of the marks in Paper I against the marks in Paper II, and draw a line of regression.

Explain how this line will indicate if the candidates found one paper more difficult than the other.

State two other ways of comparing the difficulty of the two papers. (*J.M.B.*)

7. A general knowledge test consisting of a hundred questions was given to fifteen boys of different ages with results as follows:

Boy	Age years	Age months	No. of questions correct	Boy	Age years	Age months	No. of questions correct
A	11	7	18	I	14	7	28
B	11	1	19	J	15	6	25
C	12	8	23	K	15	9	33
D	12	0	26	L	15	7	31
E	13	5	25	M	16	11	36
F	13	6	31	N	17	1	32
G	14	9	24	O	16	10	40
H	15	3	32				

Plot a scatter diagram of the number of questions correct (y) against the age in months (x), using one inch to represent 10 correct questions on the y-axis and one inch to represent 10 months on the x-axis. Sketch in the line of regression of y on x. State, with reasons, which boy deserves the prize for the best performance, taking age into consideration. (*J.M.B.*)

Equation of a Straight Line

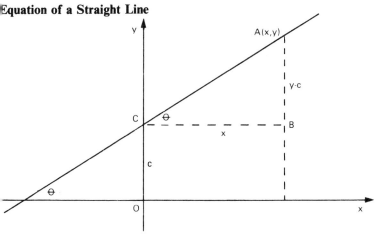

Let CA be a straight line making an angle θ with the x-axis and intersecting the y-axis at C, where $OC = c$. Let A be any point (x,y) on CA.

Then $AB = y-c$; $CB = x$, and
$$\frac{y-c}{x} = \tan\theta$$
If CA is a straight line, $\tan\theta$ is a constant. Denote $\tan\theta$ by m. Then
$$\frac{y-c}{x} = m$$
or,
$$y = mx+c$$
and this is the equation of a straight line. Note that the terms in x and y are of the first degree. It makes no difference what the constants m and c are called, and it is quite common to write this linear equation in the form
$$y = a_0+a_1 x$$

This is the best way to express y in terms of x. If we wished to express x in terms of y it is better to write the equation as
$$x = a_0+a_1 y$$

Of course, each of the constants a_0 and a_1 has a different value in the two equations.

To Find the Values of a_0 and a_1
If we consider Example 2, page 213, and wish to express P in terms of W we can find the values of the constants quite easily from the graph. The equation will be
$$P = a_0+a_1 W$$
and from the graph we note that when $W = 10$ newtons, $P = 3$ newtons and when $W = 50$ N, $P = 11$ N.
Substituting in the equation, we have
$$3 = a_0+10a_1$$
$$11 = a_0+50a_1$$
and solving for a_0 and a_1 we have
$$a_0 = 1 \text{ and } a_1 = 0.2$$
So P can be expressed in terms of W as
$$P = 1+0.2W$$
Notice the two pairs of values for P and W were taken at extreme ends of the line. This gives more accurate values for a_0 and a_1.

Regression Lines
In the above example, we expressed P in terms of W and this is the regression line of P on W.

In the example on page 219, we have two regression lines, and their equations can be found in a similar way. Let us first find the equation of the regression line of English marks on French marks. Write it as
$$E = a_0+a_1 F$$

It is the line PMQ and from the graph we get the two sets of values $E = 16$, $F = 20$; and $E = 72$, $F = 108$.

Substituting,
$$16 = a_0 + 20a_1$$
$$72 = a_0 + 108a_1$$

and solving, $a_0 = \dfrac{36}{11}$, and $a_1 = \dfrac{7}{11}$

Therefore,
$$E = \frac{36}{11} + \frac{7}{11}F$$

To find the equation of the regression line of French marks on English marks, we write the equation as
$$F = a_0 + a_1 E$$
This time we want the equation of the line RMS and, taking two convenient points, we have $F = 24$, $E = 8$; and $F = 94$, $E = 72$. Substituting and solving, we have

$$F = \frac{61}{4} + \frac{35}{32}E$$

Exercise 16.4

1. In the example on page 222, calculate the equation of the regression line of (i) Physics on Mathematics, (ii) Mathematics on Physics.

2. The following table gives Y, the total amount of gas made (in thousand million cubic metres) and X, the amount of oil used for making gas (in thousand tonnes) for each of ten consecutive months:

X	11·6	9·6	9·0	6·8	5·9	6·4	7·8	8·6	13·2	11·1
Y	11·3	9·9	9·2	8·2	7·7	7·7	8·6	9·3	10·9	11·2

   Plot this data and draw the line of regression of X on Y. From this graph, find a suitable equation for this line. (L)

3. The following table gives the cost $£C$ of transporting by air cars of length L metres:

L	4	$4\frac{1}{3}$	$4\frac{2}{3}$	5	$5\frac{1}{3}$	6
C	65	85	115	145	185	220

   Find by a graphical method, a suitable formula giving C in terms of L and use your formula to suggest a value of C when $L = 5\cdot2$. (L)

4. The following table gives X, the number of radio sets (in thousands) and Y, the number of television sets (in thousands) sold in the home market for each of nine months from September, 1958 to May, 1959.

X	137	148	130	100	93	83	87	119	129
Y	269	353	345	239	192	168	152	177	166

Plot this data and draw the line of regression of X on Y.
From your graph, find a suitable equation for this line.   (L)

Relationships, Reducible to the Linear Form

Relationships that are not linear can often be reduced to the linear form. For example, the relationship

$$y = a_o + a_1 x^2$$

gives a parabolic graph but, if X is written for x^2, the equation becomes linear.

$$y = a_o + a_1 X$$

Example

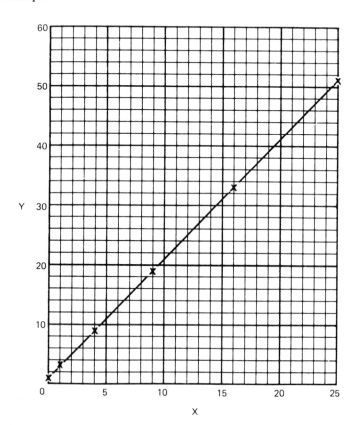

It is believed that x and y are connected by the equation $y = a_0 + a_1 x^2$. Test whether this is true for the following values of x and y, and calculate the equation connecting x and y.

$$x: \quad 0 \quad 1 \quad 2 \quad 3 \quad 4 \quad 5$$
$$y: \quad 1 \quad 3 \quad 9 \quad 19 \quad 33 \quad 51$$

Rewrite the values as follows.

$$X = x^2: \quad 0 \quad 1 \quad 4 \quad 9 \quad 16 \quad 25$$
$$y: \quad 1 \quad 3 \quad 9 \quad 19 \quad 33 \quad 51$$

Plot X against y. The points lie on a straight line (see graph on page 232).
From the graph, we get the points $X = 0$, $y = 1$; and $X = 25$, $y = 51$.

Substituting in the equation
$$y = a_0 + a_1 X$$
and solving, we get
$$y = 1 + 2X$$
or, in the original variables,
$$y = 1 + 2x^2$$

Example
An experiment was conducted to find the relation between two quantities x and y. The results obtained were as follows:

$$x: \quad 2 \quad 3 \quad 4 \quad 5 \quad 10 \quad 15 \quad 20$$
$$y: \quad 0{\cdot}7 \quad 0{\cdot}8 \quad 0{\cdot}9 \quad 1{\cdot}3 \quad 1{\cdot}3 \quad 1{\cdot}7 \quad 1{\cdot}4$$

Change to a new pair of variates X and Y, where $X = \dfrac{1}{x}$ and $Y = \dfrac{1}{y}$.

Plot a scatter diagram for Y against X and draw, as accurately as you can, the line of regression of Y on X.

Find an equation for y in terms of x.

Writing the values of X and Y, we have:

$$X = \frac{1}{x} \quad 0{\cdot}5 \quad 0{\cdot}33 \quad 0{\cdot}25 \quad 0{\cdot}2 \quad 0{\cdot}1 \quad 0{\cdot}07 \quad 0{\cdot}05$$

$$Y = \frac{1}{y} \quad 1{\cdot}42 \quad 1{\cdot}25 \quad 1{\cdot}11 \quad 0{\cdot}77 \quad 0{\cdot}77 \quad 0{\cdot}59 \quad 0{\cdot}71$$

Calculating the means of X and Y and the upper and lower array means, we have, $\overline{X} = 0{\cdot}214$; $\overline{Y} = 0{\cdot}95$; $\overline{X}_1 = 0{\cdot}36$; $\overline{Y}_1 = 1{\cdot}26$; $\overline{X}_2 = 0{\cdot}105$; $\overline{Y}_2 = 0{\cdot}71$.

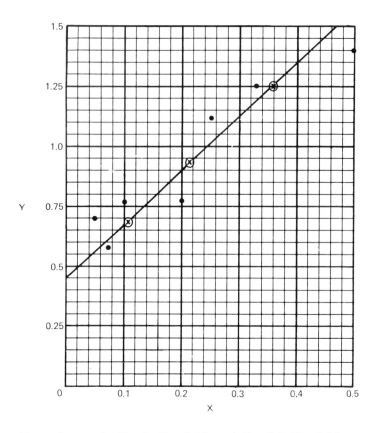

From the graph, $X = 0$, $Y = 0.45$; and $X = 0.4$, $Y = 1.35$.
Substituting in $\qquad Y = a_0 + a_1 X$
and solving, we have $\qquad Y = 0.45 + 2.25\,X$
or, in the original units,

$$\frac{1}{y} = 0.45 + 2.25\frac{1}{x}$$

Exercise 16.5

1. The following table gives corresponding values of two variables x and y:

y	5·3	11	22·3	35·2	53·1	76
x	1	2	3	4	5	6

It is suspected that there is a relation between x and y of the form $y = a + bx^2$, where a and b are constants.
(i) Plot the values of y against x^2.
(ii) Draw the line of best fit.
(iii) Find suitable values of a and b. $\qquad$ (L)

234

2. The following table gives corresponding values of two variables, W and d.

d:	2	3	4	5	6	7
W:	62·0	53·1	44·9	40·0	27·5	11·0

It is suspected that there is a relation between W and d of the form $W = a + bd^2$, where a and b are constants.

(i) Plot the values of W against d^2.

(ii) By a graphical method, find suitable values of a and b. **(L)**

3. The following table gives corresponding values of two variables, x and y.

x:	1	0·4	0·25	0·17	0·125	0·12	0·1
Y:	30	25	23·6	20·8	15·1	14·9	13·8

It is suspected that there is a relation between x and y of the form $y = a + \dfrac{b}{x}$, where a and b are constants.

(i) Plot the values of y against $\dfrac{1}{x}$.

(ii) By a graphical method find suitable values of a and b. **(L)**

4. In the following table M stands for the number of motor-car and motor-cycle licences issued per year, and P for the passenger receipts for public road passenger transport.

Year:	1952	1953	1954	1955	1956	1957	1958
M(100 000's)	35	38	42	48	52	57	61
P(£10 millions)	25	26	26	28	29	30	30

Year:	1959	1960	1961
M	67	74	78
P	31	32	33

Draw a scatter diagram for M and P and draw the line of best fit.

Find a relation between P and M of the form
$$P = aM + b \qquad \text{(A.E.B.)}$$

5. An experiment was conducted to find the relation between two quantities x and y. The results obtained were as follows:

x	2	3	4	5	10	15	20
y	1·4	1·2	1·1	0·8	0·8	0·6	0·7

Change to a new pair of variables X and Y where $X = \dfrac{1}{x}$ and $Y = y$.

Plot a scatter diagram for Y against X and draw as accurately as you can the line of regression of Y on X.

Find an equation for y in terms of x. **(A.E.B.)**

6. The following table gives corresponding values of two variables x and y.

x:	0	1	$1\frac{1}{2}$	2	$2\frac{1}{2}$	3	$3\frac{1}{2}$
y:	3·2	4·8	8	11·3	15·2	20·7	27·6

Plot the values of y against x^2. Draw, as accurately as you can, a line of regression of y on x^2 and find a suitable equation giving y in terms of x^2.

Find the value of y that corresponds to $x = 1·6$, and the value of x that corresponds to $y = 24·8$.   (A.E.B.)

7. In an experiment, the following values were obtained for two variables x and y.

x:	0	0·5	1	1·5	2	2·5	3	3·5
y:	0·6	0·45	0·8	0·85	1·4	1·65	2·4	2·85

Change to a new pair of variables X and Y where $X = x^2$ and $Y = y$.

Plot a scatter diagram for Y against X and hence draw the line of regression for Y on X.

Find an equation for y in terms of x.   (A.E.B.)

8. Corresponding values of x and y are given as follows:

x	$\frac{1}{2}$	1	$1\frac{1}{2}$	2	$2\frac{1}{2}$	3
y	$\frac{2}{7}$	$\frac{1}{2}$	$\frac{2}{3}$	$\frac{4}{5}$	$\frac{10}{11}$	1

Plot the values of x/y against the values of x, and hence find the equation connecting x and y.   (J.M.B.)

9. The following pairs of values connect x and y:

x	0	0·5	1·0	1·5	2·0	2·5
y	3·5	3·6	3·9	4·4	5·1	6·0

Plot the values of y against the values of x^2 and hence find the equation connecting x and y.   (J.M.B.)

17 Correlation

Coefficient of Correlation

The statistician has to find a coefficient which can be used to measure the degree of linear correlation between two variables. The derivation of the coefficients of correlation is beyond the scope of this book, but the commonest is the product-moment correlation coefficient, which is given by:

The product-moment coefficient of correlation $=$

$$\frac{\text{the mean of the product of the deviations from the mean of each variable}}{\text{the standard deviation of the first} \times \text{the standard deviation of the second}}$$

If the correlation between two sets of numbers is perfect and also direct, (i.e. high values of one variable being related to high values of the other variable), the coefficient will be $+1$. If the correlation is perfect but inverse, (i.e. high values of one variable being related to low values of the other variable), the coefficient will be -1, and the smaller the numerical value of the coefficient, the less the degree of correlation between the variables.

The numerical calculation of the product-moment coefficient of correlation can be simplified, for if x is the deviation from the mean of one of the first variables, and y is the deviation from the mean of the corresponding second variable, then

product-moment coefficient of correlation $=$

$$\frac{\dfrac{\text{Sum of terms like } xy}{n}}{\sqrt{\dfrac{\text{Sum of terms like } x^2}{n}} \times \sqrt{\dfrac{\text{Sum of terms like } y^2}{n}}}$$

$$= \frac{\text{Sum of terms like } xy}{\sqrt{\text{Sum of terms like } x^2} \times \sqrt{\text{Sum of terms like } y^2}}$$

Example
The following table gives the marks of six candidates in two tests **A** and **B**. Calculate the product-moment coefficient of correlation.

Test A	1, 3, 4, 6, 7, 9.			Test B	5, 3, 4, 10, 6, 8.		
Marks for test A	Deviation from mean, x	x^2		Marks for test B	Deviation from mean, y	y^2	xy
1	-4	16		5	-1	1	4
3	-2	4		3	-3	9	6
4	-1	1		4	-2	4	2
6	1	1		10	4	16	4
7	2	4		6	0	0	0
9	4	16		8	2	4	8
Mean = 5		Total 42		Mean = 6		Total 34	24

$$\text{Product-moment coefficient of correlation} = \frac{24}{\sqrt{42 \times 34}} = 0.64$$

Exercise 17.1

1. The following table shows the examination marks of eight students in Algebra and Geometry. Calculate the coefficient of correlation.

Algebra	10	24	30	35	48	59	68	70
Geometry	31	43	48	59	72	73	90	96

2. In twelve areas the percentage of dwellings overcrowded, X, and the infant mortality, Y, were as follows. Calculate the coefficient of correlation.

X	12	34	11	40	13	8	19	3	17	26	14	7
Y	125	149	127	158	130	76	128	110	130	145	110	64

3. The mass and height of 10 boys were measured and the results were as follows:

Height (cm)	130	132	135	137	140	141	145	147	149	164
Mass (kg)	32	32	38	38	38	44	51	44	51	77

   Calculate the coefficient of correlation.

4. The first of the figures in each of the following brackets shows the age in years and months, and the second shows the height in cm, of the 10 boys in a class. For example the bracket (16.5, 175) means that one boy is 16 yr. 5 m. in age and 175 cm in height.

   (16.5, 175), (15.4, 172·5), (16.7, 162·5), (16.0, 170), (15.1, 167·5), (17.3, 175), (14.9, 160), (16.11, 162·5), (14.4, 147·5), (15.3, 157·5).

   Calculate the coefficient of correlation between age and height and comment on your result.

   (You may assume the following numbers: the mean age is $15.9\frac{1}{2}$;

the mean height is 165. The sum of the squared deviations of the ages from $15.9\frac{1}{2}$ is $1248\frac{1}{2}$; that of the heights from 165 is 687·5.)

(CA.)

5. The table shows the index figures for production and for price of an article over ten consecutive years. Calculate the coefficient of correlation. Comment on your result.

Year	1	2	3	4	5	6	7	8	9	10
Production	92	96	103	108	109	108	96	103	109	103
Price	109	111	94	93	89	84	100	106	87	97

(CA.)

Rank Correlations

When the coefficient of correlation is calculated as above, it is necessary that the data should be definitely measured. But in some types of inquiry we may have to deal with qualities which are not expressible as numbers of units of an objective kind. For example, it is difficult to measure an individual's 'ability' in any particular skill, such as ability in music or colour selection, but often it is possible to place a group of individuals in order of skill. They are then said to be *ranked*. We can also rank individuals in some property that it is possible to measure without actually measuring the property; for example, we can rank a group of individuals with respect to height without measuring the height of individuals.

If we have a set of individuals ranked according to two different qualities, it is natural to inquire whether the ranks can be made to give us the measure of correlation between the two qualities. The rank coefficient of correlation is deduced from the previous correlation coefficient, and it is known as Spearman's rank correlation coefficient. It is denoted by ρ (rho), and

$$\rho = 1 - \frac{6 \times (\text{sum of the squares of the differences in rank})}{n(n^2 - 1)}$$

where n is the number of individuals.

Example

The ranking of twelve students in Mathematics and Music are as follows:

Mathematics	1	2	3	4	5	6	7	8	9	10	11	12
Music	6	4	5	1	2	8	7	10	12	11	3	9

What is the coefficient of rank correlation?

The differences, d, are (mathematical rank minus musical rank):

$$-5 \ -2 \ -2 \ 3 \ 3 \ -2 \ 0 \ -2 \ -3 \ -1 \ 8 \ 3$$

(These add up to zero, as they should.)

239

The squares of d are

$$25 \quad 4 \quad 4 \quad 9 \quad 9 \quad 4 \quad 0 \quad 4 \quad 9 \quad 1 \quad 64 \quad 9$$

which add up to 142.

Hence the rank correlation coefficient $\rho = 1 - \dfrac{6 \times 142}{12(144 - 1)}$

$$= 0{\cdot}50.$$

The rank correlation coefficient also varies from $+1$ to -1.

Equal Ranks

In some cases it may be found necessary to rank two or more individuals as equal. In this case it is customary to give each individual an average rank. Thus if two individuals are ranked equal at sixth place, they are each given the rank $\dfrac{6+7}{2}$, that is $6\frac{1}{2}$, while if three are ranked equal at sixth place, they are given the rank $\dfrac{6+7+8}{3}$, that is 7. This is not mathematically correct, but it is a satisfactory working rule if only a few items are ranked equal. The sum of the deviations still equals zero.

Example

The percentage marks for ten candidates in each of two tests were as follows:

| Test 1 | 27 | 39 | 47 | 43 | 47 | 55 | 36 | 30 | 37 | 48 |
| Test 2 | 41 | 43 | 67 | 65 | 76 | 85 | 62 | 58 | 45 | 57 |

Calculate the rank correlation coefficient.

The work is best arranged as follows:

Marks for Test 1	Rank for Test 1	Marks for Test 2	Rank for Test 2	Difference in rank d	d^2
27	10	41	10	0	0
39	6	43	9	-3	9
47	$3\frac{1}{2}$	67	3	$\frac{1}{2}$	$\frac{1}{4}$
43	5	65	4	1	1
47	$3\frac{1}{2}$	76	2	$1\frac{1}{2}$	$2\frac{1}{4}$
55	1	85	1	0	0
36	8	62	5	3	9
30	9	58	6	3	9
37	7	45	8	-1	1
48	2	57	7	-5	25
			Totals	0	$56\frac{1}{2}$

Therefore,

$$\rho = 1 - \frac{6 \times 56\frac{1}{2}}{10 \times 99} = 0{\cdot}66.$$

The drawback of the rank correlation coefficient is that no allowance is made for the difference in values of two items. Thus two values 100 and 99 would be ranked 1 and 2, and so would two items whose values are 100 and 1.

Exercise 17.2

1. Calculate the rank correlation coefficient for the examples 1 to 5 of Exercise 17·1.

2. The following table gives the marks obtained by each of ten candidates in Physics and Chemistry. Calculate the coefficient of rank correlation between the two subjects. State what significance you can attach to the result.

Physics	70	67	52	64	44	93	66	85	75	70
Chemistry	55	56	47	50	32	66	43	72	43	40

3. Three judges in a beauty competition ranked the contestants as follows:

Contestant	A	B	C	D	E	F	G	H
Judge 1	4	3	1	6	7	8	2	5
Judge 2	6	5	4	7	1	2	3	8
Judge 3	7	6	5	4	8	3	1	2

   What is the rank correlation coefficient between either pair of judges?

   If you were the referee how would you finally rank the contestants?

4. The following is the ranking in the Cricket Championship Table for the seasons 1968 and 1969. Calculate the coefficient of correlation between the ranks.

County	Year 68	69	County	Year 68	69	County	Year 68	69
Glam.	1	3	Sussex	7	17	Yorks.	13	1
Glos.	2	16	Notts.	8	4	Leics.	14	9
Surrey	3	15	Northants.	9	13	Lancs.	15	6
Warwicks.	4	11	Kent	10	2	Derby	16	8
Hants.	5	5	Middx.	11	10	Somerset	17	12
Essex	6	14	Worcs.	12	7			

241

5. Calculate the rank correlation coefficient for the following sets of ranks, and comment on the results:

   (a) **A** 1 2 3 4 5 6
       B 1 2 3 4 5 6

   (b) **A** 1 2 3 4 5 6
       B 6 5 4 3 2 1

   (c) **A** 1 2 3 4 5 6
       B 4 1 6 3 5 2

The Interpretation of the Correlation Coefficient

We must be careful not to misuse the correlation coefficient. It is only a measure of the degree to which large and small values of one variable are associated with large and small values of another variable. It is wrong to infer that a direct causal relation exists between two variables because they have a high correlation coefficient. A high coefficient of correlation will be obtained with any two sets of numbers that show a similar trend. But, in particular cases, if a high degree of correlation exists, it may be correct to infer that a causal relation exists; e.g. the number of persons suffering from a disease and the extent of over-crowding. Another possible inference is that the values of both variables depend upon the values of a third variable, but correlation does not necessarily mean causation.

It is also wrong to infer that no relationship exists between two sets of figures with a low correlation coefficient. The correlation coefficient shows a linear relationship, and not, say, a parabolic relationship. Thus the product-moment correlation coefficient for the following set of measurements is 0, yet if they are plotted on a graph, they are seen to lie on a parabola, $y = x^2 - 5x + 8$.

Measurement **A** 0 1 2 3 4 5
Measurement **B** 8 4 2 2 4 8

Exercise 17.3

1. Explain what is meant by correlation and give examples of data where (a) direct correlation, (b) indirect correlation might be expected.

   The following table gives the rainfall and sunshine for each month of the year 1946 expressed as a percentage of the mean for the preceding 25 years. Calculate the coefficient of ranked correlation between rainfall and sunshine.

	Jan.	Feb.	Mar.	Apr.	May	June
Rainfall	114	132	56	70	134	139
Sunshine	99	114	93	126	105	88

	July	Aug.	Sept.	Oct.	Nov.	Dec.	
Rainfall	108	167	193	37	178	102	
Sunshine	103	84	77	76	63	149	(L)

2. Explain what is meant by correlation.

There are ten finalists in a competition for which there are two judges **X** and **Y**. The following table gives the order in which **X** and **Y** place the competitors. Calculate the rank correlation coefficient and state any conclusions that you deduce from it.

Competitors	A	B	C	D	E	F	G	H	I	J
Ranking of **X**	6	10	2	7	1	4	5	9	8	3
Ranking of **Y**	2	7	8	10	6	1	9	5	3	4

(L)

3. If r is the coefficient of correlation, discuss the meaning of the following:

$$(a) \; r = 0, \; (b) \; r = -1, \; (c) \; -1 < r < +1.$$

The number of television licences (in thousands) and the number of admissions to cinemas (in millions) for each quarter from January 1957 to March 1959 are given in the following table:

Licences	20586	21339	21999	22942	23983	24601
Admissions	264·0	226·2	235·4	189·7	199·3	191·4

Licences	25064	26201	27452
Admissions	199·5	163·8	162·9

Calculate the coefficient of rank correlation between these items.

(L)

4. Ten pupils sat two mathematics tests. The marks in the two tests were as follows:

Pupil	A	B	C	D	E	F	G	H	I	J
Test 1	15	17	15	6	8	14	7	6	3	12
Test 2	14	17	14	8	5	14	5	8	4	10

Calculate the coefficient of rank correlation between the two sets of marks. (A.E.B.)

5. What does the correlation coefficient measure?

The table gives the marks of 10 candidates in two examinations as a deviation from 50 marks.

Candidate	A	B	C	D	E	F	G	H
Examination 1	+27	−15	+19	+ 9	+ 5	− 8	+36	+ 5
Examination 2	+ 7	−10	+ 7	−10	+16	−38	+15	−17

I	J
+11	−14
−35	−10

Calculate the coefficient of rank correlation between the two sets of marks. *(A.E.B.)*

6. The prices in pence of a certain product in six self-service stores on 6 December 1975 and 7 February 1976 are given below.

Store	A	B	C	D	E	F
Price in pence 6 December 1975	20	18	17	21	15	14
Price in pence 7 February 1976	21	22	18	23	20	19

Calculate Spearman's coefficient of rank correlation between the prices on the two dates.

Comment on the result. *(A.E.B.)*

7. Seven electric light bulbs of the same size but from different firms were purchased and tested for the number of hours illumination they provided. The results are tabulated below.

Firm	Price (p)	Hours of Illumination
A	$8\frac{1}{2}$	1090
B	10	980
C	$9\frac{1}{2}$	1150
D	11	1020
E	8	960
F	$9\frac{1}{2}$	910
G	9	900

Calculate a coefficient of rank correlation and comment as to whether price is a good guide to performance. *(A.E.B.)*

8. (a) Which of the following coefficients express
   (i) perfect correlation,
   (ii) no correlation?
   $$-1.00, \ -0.50, \ 0.00, \ 0.50, \ 1.00.$$

   (b) A group of pupils sat a trial examination in May and the official examination in June. The marks obtained by the pupils in both examinations were as follows:

   Trial marks: 10 14 23 36 36 44 45 55 56 60 72 85
   Official marks: 18 7 28 40 42 45 45 45 55 50 73 68

   Rank the pupils in both examinations and calculate Spearman's rank correlation coefficient. *(A.E.B.)*

9. The weekly takings and the weekly profits for five branch shops of a firm are set out below.

Shop	1	2	3	4	5
Takings £	4000	6200	3600	5100	5000
Profits £	400	1100	450	750	800

Calculate the coefficient of rank correlation between the takings and the profits.

Find also the value of this coefficient if the weekly takings for each shop increase by 10% and the profit in each case rises by £200.

18 Probability or Chance

Probability

Example

What is the probability, or chance, of drawing a spade from a well-shuffled pack of cards?

There are 52 possible cards, and of these 13 are spades. Therefore the chance of drawing a spade is 13 out of 52, or 1 out of 4. This is written $\frac{1}{4}$.

The chance of not drawing a spade is 39 out of 52, or $\frac{3}{4}$.

Definition. If an event can happen in 'a' ways out of 'n' equally possible ways, then the probability of the event happening is $\frac{a}{n}$. Similarly if the event can fail to happen in 'b' ways out of 'n' equally possible ways, the probability of the event not happening is $\frac{b}{n}$. Obviously $a+b = n$.

The letter p is generally used to denote the probability of an event happening, and the letter q the probability of its failing.

Thus $p = \frac{a}{n}$, and $q = \frac{b}{n}$.

Now $p+q = \frac{a}{n}+\frac{b}{n} = \frac{a+b}{n} = 1$.

Since the event must either happen or not happen, the measure of certainty is 1.

Example

What is the probability of obtaining two heads when two pennies are tossed together?

If the first coin turns up heads, the second coin may be either heads or tails. This gives us two possibilities. If the first coin turns up tails, the second coin may again be either heads or tails. This gives us two more possibilities. Therefore the total number of possibilities is 4, and only once do two heads appear.

Therefore $p = \frac{1}{4}$

and $q = \frac{3}{4}$.

Exercise 18.1

1. If the probability of an event happening is (a) $\frac{1}{4}$, (b) $\frac{1}{3}$, (c) $\frac{5}{8}$, (d) $\frac{1}{2}$, (e) $\frac{7}{8}$, (f) $\frac{3}{8}$, (g) $\frac{3}{4}$, what is the probability of its failing?

2. What are the chances of events (a) to (i) happening?
    (a) Drawing a red ball from a bag containing 5 red balls, and 20 blue balls?
    (b) Drawing a king out of a pack of 52 cards containing 4 kings?
    (c) Drawing a picture card out of a pack of 52 cards containing 12 picture cards?
    (d) Throwing a number greater than 4 with a dice whose faces are numbered 1 to 6?
    (e) Selecting a boy from a class containing 12 girls and 20 boys, if the pupil is selected at random?
    (f) Obtaining one head and one tail if two coins are spun simultaneously?
    (g) Throwing either a four, five, or a six, with a die whose faces are numbered 1 to 6?
    (h) Winning a prize in a raffle if you have bought 10 of the 400 tickets sold?
    (i) Picking an odd number out of the numbers 1 to 11?

3. A bench can seat 9 boys. What is the chance of any one boy sitting (a) in the middle, (b) on the end of the bench?

4. The probability of an event happening is $\frac{4}{15}$. What is the probability of it not happening?

5. A person's chance of winning a raffle is $\frac{1}{36}$. If he holds 10 tickets, how many tickets have been sold?

6. In a bag there are 6 red balls, 5 white balls, 10 blue balls, and 9 black balls. If one ball is chosen at random, what is the probability of it being (a) red, (b) white, (c) blue, (d) black?

7. Two dice are thrown together. What is the probability of the total score being (a) 2, (b) 4, (c) 5, (d) 10, (e) 12, (f) 14?

8. If a letter is chosen at random from the word (a) unearthly, (b) hopeful, (c) merchant, (d) jouster, what is the chance of it being a vowel?

9. A bag contains b blue balls and r red balls. What is the probability of selecting (a) a blue ball, (b) a red ball?

10. A person has a bag containing 10 oranges and 7 apples. He selects a fruit at random. What is the chance of it being an apple?

If his first selection was an apple, what is the chance of his second selection being an orange?

11. Find the chance of throwing a 9 in a single throw with two dice.

Mutually Exclusive Events
Definition. Two events are said to be mutually exclusive when both cannot happen at the same time.

For example, if a coin is tossed, the appearance of a head excludes the possibility of the appearance of a tail. Therefore the appearance of a head or the appearance of a tail are mutually exclusive events.

Theorems on Probability
There are three useful theorems on probability.

Theorem 1   The theorem of Total Probability, sometimes known as the theorem of 'either...or'.

Example
What is the probability of drawing either a red numbered card or a black picture card from a set of 52 playing cards?

The number of ways of selecting a red numbered card is 20, and of selecting a black picture card is 6. Therefore the number of ways of selecting either a red numbered card or a black picture card is $20+6 = 26$. There are 52 possible selections. Therefore the probability is $\dfrac{26}{52}$, or $\frac{1}{2}$.

General proof of theorem
If two events are mutually exclusive, and the probability of the first event happening is p_1, and of the second event p_2, then the probability of *either* the first event *or* the second event happening is p_1+p_2.

For if the first event can happen in a_1 ways, and the second event in a_2 ways, then the number of ways in which either event can happen is a_1+a_2. If the total number of possibilities is n, then by definition, the probability of either the first or the second event happening is $\dfrac{a_1+a_2}{n}$.

Now $p_1 = \dfrac{a_1}{n}$, and $p_2 = \dfrac{a_2}{n}$. Therefore if p is the probability of either event happening

$$p = \frac{a_1+a_2}{n} = \frac{a_1}{n} + \frac{a_2}{n} = p_1 + p_2.$$

In a similar way the theorem can be extended to three or more mutually exclusive events.

Exercise 18.2

What is the probability of either one or the other event happening in questions 1 to 4?

1. Drawing a picture card or an ace from a pack in a single trial?
2. A five or a three turning up on a six-sided die?
3. Selecting either a red or a blue ball from a bag containing 4 red balls, 6 blue balls, and 13 white balls?
4. Drawing a card numbered either 4, 5, 6, or 7, from a pack of 52 playing cards?
5. The probability of an event happening is $\frac{1}{3}$, and of another mutually exclusive event $\frac{3}{8}$. What is the probability of either event happening?
6. A vase contains 4 red roses, 3 pink roses, 5 yellow roses, and 8 white roses. What are the chances of a blind person selecting either a red or a white rose?
7. Twenty cards are marked from 1 to 20, and one is drawn at random. What is the chance of it being a multiple of either 2 or 5?

Theorem 2   The theorem of Compound Probability, sometimes known as the theorem of 'both...and'.

Example

What is the chance of forecasting the results of two games of football?

The first game can have three results, a home win, a home loss, or a draw; and the second game can also have three results. Each of the results of the second game can be combined with each of the results of the first game, and this gives us nine possible results. Only one result can be correct. Therefore the probability of success is $\frac{1}{9}$.

General proof of theorem

If the probability of one event happening is p_1, and of another independent event p_2, then the probability of *both* the first *and* second event happening is $p_1 \times p_2$.

For if the first event can happen in n_1 ways, of which a_1 are successful, and the second event in n_2 ways, of which a_2 are successful, then since each successful event in the first case can be combined with each successful event in the second case, the total number of successful possibilities in both cases is $a_1 \times a_2$. Similarly the total number of possible cases is $n_1 \times n_2$.

Then, by definition, the probability of *both* independent events happening is

$$\frac{a_1 \times a_2}{n_1 \times n_2} = \frac{a_1}{n_1} \times \frac{a_2}{n_2} = p_1 \times p_2.$$

In a similar way the theorem can be extended for three or more events.

Example
Prove that the sum of the probabilities of all possibilities in two independent events amounts to certainty.

Let the probabilities of success and failure in the first event be p_1 and q_1, and in the second event p_2 and q_2. Then,
the chance of success in the first event and success in the second event is $p_1 \times p_2$,
the chance of success in the first event and failure in the second event is $p_1 \times q_2$,
the chance of failure in the first event and success in the second event is $q_1 \times p_2$,
the chance of failure in the first event and failure in the second event is $q_1 \times q_2$.

These are all the possibilities, and the sum of the probabilities is
$$\begin{aligned}
& p_1 \times p_2 + p_1 \times q_2 + q_1 \times p_2 + q_1 \times q_2 \\
&= p_1(p_2 + q_2) + q_1(p_2 + q_2) \\
&= p_1 + q_1 \quad \text{(since } p_2 + q_2 = 1) \\
&= 1. \quad \text{(since } p_1 + q_1 = 1).
\end{aligned}$$

Exercise 18.3
What is the probability of both events happening in questions 1 to 4?
1. Drawing an ace from one pack of playing cards, and a king from a second pack?
2. Drawing a picture card from one pack of cards, and a numbered card from a second pack?
3. Tossing a coin and a die and obtaining a head and a six?
4. Forecasting correctly the results of 12 football games?
5. The probabilities of each of two independent events **A** and **B** happening is p_1 and p_2. Calculate the probability of
   (*a*) both **A** and **B** happening.
   (*b*) **A** happening and **B** failing.
   (*c*) **B** failing and **A** failing.
   (*d*) **A** failing and **B** happening.
   if (i) $p_1 = \frac{1}{2}$, $p_2 = \frac{1}{3}$; (ii) $p_1 = \frac{1}{4}$, $p_2 = \frac{1}{8}$; (iii) $p_1 = \frac{1}{3}$, $p_2 = \frac{1}{8}$; (iv) $p_1 = \frac{3}{8}$, $p_2 = \frac{3}{4}$; (v) $p_1 = \frac{5}{8}$, $p_2 = \frac{7}{8}$.
6. If the chances of success in four independent events **A, B, C, D**, are p_1, p_2, p_3, p_4, respectively, what is the chance that
   (*a*) **A** occurs, and **B, C, D**, do not?
   (*b*) **A** and **B** occur, and **C** and **D** do not?

(c) **A**, **B**, **C**, occur, and **D** does not?

(d) **A**, **B**, **C**, and **D** occur?

7. The chance of any particular day in December being frosty is $\frac{1}{3}$. What is the chance of it being frosty from Dec. 24th to Dec. 31st, inclusive?

8. What is the chance of drawing two white balls in succession from a bag containing 7 red and 5 white balls, the ball drawn first not being replaced?

9. The letters of the alphabet are written on cards and the cards are thoroughly shuffled. What is the probability of drawing the cards marked C, A, R, D, S, on them in that order?

Theorem 3   The 'At least one...' theorem.

Example

Calculate the probability of obtaining at least one six in a throw of three dice.

The probability of obtaining a six with one die is $\frac{1}{6}$, and the probability of not obtaining a six is $\frac{5}{6}$. Similarly with the second and third dice. Therefore the chance of not obtaining a six with the three dice is $(\frac{5}{6})^3$.

Now, we must either obtain no sixes or at least one six.

Therefore, the probability of obtaining at least one six is

$$1 - \left(\frac{5}{6}\right)^3 = \frac{9}{216}.$$

General proof of the theorem

If the probabilities of n independent events are $p_1, p_2, p_3, \ldots p_n$, then the probability of at *least one* event happening is

$$1 - (1-p_1)(1-p_2)(1-p_3)\ldots(1-p_n).$$

The chance of the first event not happening is $q_1 = 1 - p_1$, and of the second event $q_2 = 1 - p_2$, and so on. Therefore by theorem 2 the chance of not one of the events happening is $q_1 q_2 q_3 \ldots q_n$.

Now it is certain that either no event happens or *at least one* event happens. Therefore the probability of at least one event happening is

$$1 - q_1 q_2 q_3 \ldots q_n$$
$$= 1 - (1-p_1)(1-p_2)(1-p_3)\ldots(1-p_n).$$

Empirical definition of probability

So far we have assumed that the probability of an event can be calculated from the nature of the problem. Thus, when we say the probability of throwing a six with a die is $\frac{1}{6}$, we make the assumption that the die is not biased. If the die is biased we cannot calculate the probability of throwing a six, but if we conduct a certain number of trials

251

and count the number of times we throw a six, we can express the probability as

$$\frac{\text{Number of times a six is thrown,}}{\text{Total number of throws}}$$

and the greater the number of throws the more exact the estimate of the probability.

This 'empirical' definition of probability is used in the first question of the next exercise.

Exercise 18.4

1. At practice one boy has broken the school record for the 100 m race once in three attempts, a second boy twice in five attempts, and a third boy once in four attempts. If these results are taken as a measure of their probability of breaking the school record, what is the probability the school record will be broken on the day of the school sports?

2. One card is selected from each of four packs. What is the probability that at least one of the selections is an ace?

3. The following table gives the number of boys and girls in each form in a school. A pupil is selected at random from each form. What is the probability of at least one girl being selected?

Form	Boys	Girls
1	20	10
2	23	9
3	25	6
4	27	4
5	30	0

4. A man holds 1, 2, 4, and 8, tickets in four raffles in which 250, 300, 400, and 600 tickets, respectively, have been sold. What is his chance of winning at least one prize?

5. One bag contains 4 white balls and 10 black balls, and another bag contains 6 white balls and 18 blue balls. If a ball is selected from each bag what is the chance of at least one of them being white?

6. Three dice are thrown. What is the probability of at least one of the numbers turning up being greater than 4?

7. A man forecasts the results of four football games. What is the probability of his giving a correct forecast in at least one game?

8. A man is selected for interview for three separate posts. At the first interview there are five candidates; at the second four candidates;

and at the third six candidates. If the chances of selection of the candidates are equally likely, what is the chance of the man obtaining at least one of the posts?

9. In a game for two players a turn consists of throwing a die either once or twice; once if the score obtained is less than six; twice if the score at the first throw is six. The score for the turn is in the first case the score of the single throw, and in the second case the total score of the two throws.

Obtain the probabilities of a player
(a) scoring more than 9 in a single turn,
(b) scoring a total of more than 20 in two successive turns,
(c) obtaining equal scores in two successive turns. (*C.A.*)

10. Two groups, each of three children, contain, respectively, two boys and one girl, and one boy and two girls. One child is drawn at random from each group. Calculate the probability that (a) both will be boys, (b) one will be a boy and the other a girl, (c) at most one boy will be selected. (*J.M.B.*)

11. Explain briefly the meaning of probability.

The table gives the number of boys and girls in each form in a school. A pupil is selected at random from each form. What is the probability
(i) of no boy being selected,
(ii) of at least one girl being selected?

Form	1	2	3	4	5
Boys	25	24	26	27	29
Girls	5	4	6	3	2

(*A.E.B.*)

12. A hundred steel rods of length 20 cm are ordered from a workshop. On delivery they are measured to the nearest hundredth of a cm, the result being shown in the table.

Length (cm)	19·98	19·99	20·00	20·01	20·02	20·03
Number	5	22	33	20	18	2

Two rods have to be selected with lengths 20·00 cm to the nearest hundredth of a cm. If two rods are chosen at random, show that the chance that they are both suitable is a little more than 1 in 10. Find also the probability that neither is suitable and the probability that one is suitable and the other is not. (*L*)

19 Permutations and Combinations

Permutations

An elementary knowledge of permutations and combinations simplifies the calculation of the probability of an event. The subject of permutations is best introduced by an example.

Example

A schoolmaster wishes to pick a form captain and a form vice-captain and has five boys who, in his opinion, equally merit the positions. In how many ways can he select the boys?

Suppose the boys are **A**, **B**, **C**, **D**, and **E**. There are five ways of selecting the captain. Suppose **A** is selected. This leaves four ways of selecting a vice-captain. Therefore the number of possible selections is $5 \times 4 = 20$.

The following table shows the arrangements:

Captain	**A A A A**	**B B B B**	**C C C C**	**D D D D**
Vice-captain	**B C D E**	**A C D E**	**A B D E**	**A B C E**

Captain	**E E E E**
Vice-captain	**A B C D**

When different arrangements are made of given items they are said to be permuted, and when they are arranged in every possible manner we have formed the permutations of the items.

In our example we have formed the permutations of five items taken two at a time.

Factorial Notation

In mathematics the product $7 \times 6 \times 5 \times 4 \times 3 \times 2 \times 1$ is written for short as $7!$ or $\lfloor 7$ and is called 'factorial' seven.

Thus, $\qquad 5! = 5 \times 4 \times 3 \times 2 \times 1$

and $\qquad 8! = 8 \times 7 \times 6 \times 5 \times 4 \times 3 \times 2 \times 1$

and so on.

The permutations of five items two at a time is written $_5P_2$, and in our example we have shown

$$_5P_2 = 5 \times 4$$

$$= \frac{5 \times 4 \times 3 \times 2 \times 1}{3 \times 2 \times 1}$$

$$= \frac{5!}{3!} = \frac{5!}{(5-2)!}$$

Similarly $_6P_4 = \frac{6!}{(6-4)!} = \frac{6!}{2!} = \frac{6 \times 5 \times 4 \times 3 \times 2 \times 1}{2 \times 1} = 360.$

By similar reasoning, the number of permutations of n items r at a time is

$$_nP_r = \frac{n!}{(n-r)!}$$

If we have three items to permute, we can select the first in three ways, the second in two ways, and the last in one way.

Therefore

$$_3P_3 = 3 \times 2 \times 1 = 3!$$

By the definition

$$_3P_3 = \frac{3!}{(3-3)!} = \frac{3!}{0!}$$

Therefore 0! must equal 1,
and

$$_nP_n = \frac{n!}{(n-n)!} = \frac{n!}{0!} = n!$$

Example
Evaluate $_{21}P_2$.

$$_{21}P_2 = \frac{21!}{(21-2)!} = \frac{21!}{19!} = 21 \times 20 = 420.$$

Example
Find n if $_nP_2 = 42$.

$$\frac{n!}{(n-2)!} = 42$$

Therefore,

$$n(n-1) = 42 \text{ [Cancel } (n-2)!]$$
$$n^2 - n = 42$$
$$n^2 - n - 42 = 0$$
$$(n-7)(n+6) = 0$$

$$n = 7 \ (n = -6 \text{ does not satisfy the problem}).$$

Example

How many three-figure numbers can be formed with the digits 1, 2, 3, 4, 5, 6, 7, 8, 9, when no digit is repeated?

We must find the number of permutations of nine things three at a time.

$$_9P_3 = \frac{9!}{6!} = 9 \times 8 \times 7 = 504$$

Exercise 19.1

1. Evaluate $_5P_3$; $_7P_2$; $_8P_8$.
2. Find n if $_nP_2 = 30$, $_nP_2 = 72$.
3. Thirteen boys can play soccer equally well in any position. How many teams can be formed?
4. How many two-figure numbers can be formed with the digits 1 to 9 when no digit is repeated?
5. How many numbers can be formed with the digits 1 to 9 when no digit is repeated?
6. In how many ways can ten different volumes be arranged on a bookshelf?
7. In how many ways can three men be accommodated each in a different room if there are ten rooms at their disposal?
8. There are five routes between two towns. How many ways are there of going from one place to another and returning by a different route?
9. There are four chairs in a room and eight persons. In how many ways can the chairs be occupied?
10. In how many ways can a captain, a vice-captain, and a secretary be selected from a club of 50 members if each member is equally capable of filling either position?
11. A chef makes out 12 different menus. In how many ways can he arrange dinners for a week if he does not use the same menu twice?
12. How many numbers ending in 3 can be made by using the digits 1, 2, 3, 4, 5, 6?

Permutations When Some Objects are Alike

Suppose we have three articles, two of which are alike, and let them be denoted by x, a, a. We can arrange these articles in the following ways: x,a,a, a,a,x, a,x,a.

If the two articles a are different, a_1 and a_2 say, the number of arrangements is $_3P_3$, and they are: xa_1a_2, xa_2a_1, a_1a_2x, a_2a_1x, a_1xa_2, a_2xa_1, i.e. the previous number of permutations multiplied by the number of ways we can permute a_1 and a_2.

Therefore, if the first number of arrangements is denoted by N, (we know in this case $N = 3$), we have

$$N(2!) = 3!$$
$$\text{or, } N = \frac{3!}{2!}$$

Similarly we can extend the reasoning as follows: to find the number of permutations of n things taken all together, when there are p alike of one kind, q alike of a second, r alike of a third, and so on.

Suppose there are N permutations, and we consider any one of these. If the p alike things are changed into p different things, then, by permuting them amongst themselves without altering the position of the rest of the things, this gives rise to $p!$ new permutations.

This will happen with each of the N permutations. Therefore the total permutations is now $N(p!)$.

Now change the q alike things into q different things. These can be permuted in $q!$ ways and each of these can be combined with the $N(p!)$ permutations.

Now we have $N(p!q!)$ permutations.

This can be done with the r alike things, and so on, so that if all the things are made different we have $N(p!q!r!\ldots)$ permutations.

But this is the number of permutations of n things when they are all different, i.e. $n!$.

Therefore $N(p!q!r!\ldots) = n!$

$$\text{or } N = \frac{n!}{p!q!r!\ldots}$$

Example

Find the number of arrangements of the letters of the word 'schoolboy'.

The total number of letters is 9, and there are 3 o's.

Therefore the number of arrangements is $\dfrac{9!}{3!} = 60480$.

Example

(i) Find the number of arrangements of the letters of the word 'consonant'.

There are 9 letters with 3 alike of one kind, and 2 alike of another kind.

Therefore the number of arrangements is $\dfrac{9!}{3!2!} = 30240$.

(ii) In how many arrangements do the three n's come together?

We may consider the three n's as forming a single letter, and then find the number of permutations of 7 letters instead of 9. Of these 7 letters two will be alike.

Therefore the number required is $\dfrac{7!}{2!} = 2520$.

(iii) How many of the arrangements begin with the three n's?

In these arrangements the first three places are filled up by the n's, and thus we merely have to find the arrangements of the remaining 6 letters two of which are alike.

Therefore, the number required is $\dfrac{6!}{2!} = 360$.

Exercise 19.2

1. How many permutations can be made from the letters of the following words, using all the letters: 'banana', 'insincere', 'proportion', 'woodwork', 'algebra', 'statistics', 'arithmetic', 'geometry'?

2. In how many ways can 5 a's, 6 b's, 3 c's, and 2 d's be arranged in a row?

3. How many arrangements can be made using all the letters of the word 'zoology'?
   How many of these arrangements will start with the letter l?
   How many will neither start with z nor finish with y?

4. A ship has 12 flags, 5 red, 3 yellow, 3 black, and 1 white. How many different signals can be made, using all the flags?

5. There are 3 copies of each of 7 different books. In how many ways can they all be arranged on a shelf?

6. How many numbers greater than 4000 can be formed by arranging the figures 1, 2, 2, 4?

7. How many numbers greater than 20 000 can be formed, using the figures 1, 1, 2, 3, 4.

8. I have 3 copies of an algebra book, 2 copies of an arithmetic book, 2 copies of a geometry book, and 1 copy of a trigonometry book. In how many ways can I lend these books to eight pupils?

9. With all the figures 0, 1, 1, 2, 3, 4, how many numbers can be formed ending in 0?

10. How many even numbers can be formed using the digits, 1, 2, 3, 4, 5, 6?

To Find the Number of Permutations of n Things Taking r at a Time When Each of the Things May be Repeated Any Number of Times

Example

How many three-figure numbers can be formed from the digits 1 to 9 when each digit may be repeated any number of times?

The first figure can be picked in 9 ways. Since repetitions are allowed the second figure can be picked in 9 ways. This gives us 9×9 ways of picking the first two figures. Similarly the third figure can be picked in 9 ways.

Therefore the total numbers formed is $9 \times 9 \times 9 = 9^3$.

Similarly, if we have n items taken r at a time and repetitions are allowed, the first item can be picked in n ways, the second in n ways, and so on. Therefore the total number of arrangements is

$$n \times n \times n \times n \times n \ldots r \text{ factors} = n^r.$$

Exercise 19.3

1. How many three-figure numbers can be formed from the digits 1, 2, 3, 4, 5, if repetitions are allowed?

2. There are three possible results to a football match. In how many ways can a forecast be made of 12 games?

3. With the figures 0 to 9 how many four-figure numbers can be formed if repetitions are allowed?

4. In how many ways could a three-letter word be made from the letters l, m, n, o, repetitions being allowed?

5. A safety lock contains 4 rings, each marked with the figures 0 to 9. How many different attempts to open the lock may be made by a person not knowing the combination?

6. How many even numbers of four digits can be formed with the numbers 0, 1, 2, 3, 5, 6, 8, if each of these digits can be repeated?

7. How many numbers of not more than four digits can be formed with the figures 1, 2, 3, if the digits can be repeated?

To Find the Total Number of Ways in Which a Selection May be Made from n Things Which are not all Different

Example

How many different groups of coins may be formed with a 50-penny, a 10-penny, three 5-penny, and two one-penny coins.

There are two choices we can make with the 50-penny coin; we can take it or reject it.

Similarly there are two choices we can make with the 10-penny coin, and each of these can be combined with each of the two choices we make with the 50-penny coin. That is, with the two coins we have 2×2 selections. There are four choices we can make with the three 5-penny coins; we can take either one 5-penny, or two 5-penny, or three 5-penny coins, or we can reject them all.

Each of these choices could be combined with the previous choices, i.e. we have, with three coins $2 \times 2 \times 4$ selections.

Similarly there are three choices we can make with the two pennies, and each of these can be combined with the previous selections. Therefore the total number of selections is

$$2 \times 2 \times 4 \times 3 = 48.$$

But this includes the case when all the coins are rejected. Therefore the total number of ways in which a selection may be made is

$$48 - 1 = 47.$$

Generally, if we have n things of which p are alike of one kind, q of another, r of another, and so on, the number of selections that can be made is

$$(p+1)(q+1)(r+1)\ldots -1$$

omitting the case where all the things are rejected.

Example

There are 12 boys and 8 girls. How many different selections could be made?

Answer: $(12+1)(8+1) - 1 = 116.$

Exercise 19.4

1. I have three similar boxes of weights, and there are five different weights in each box. How many selections can I make?
2. There are four sorts of articles: four of the first kind, three of the second kind and one each of the third and fourth kind. How many selections can be made?
3. How many selections can be made of the letters of the word 'proportion'?
4. There are a number of articles, four of which are alike, and the rest all different. 39 selections can be made. How many articles are there?
5. I have five friends. In how many ways may I invite one or more of them to dinner?

Combinations

ab and ba are two permutations of the letters ab, but if the order of the letters is not important, they are the same combination. If out of the four letters $abcd$ we pick three at a time, we have the following possible selections:

abc	abd	bcd	cda
acb	adb	bdc	cad
bac	bda	cdb	dac
bca	bad	cbd	dca
cab	dab	dbc	acd
cba	dba	dcb	adc

i.e. we have $_4P_3$ permutations. If the order of the letters in each selection is not important, each of the columns has the same letters and forms the same combination. It is obvious that the number of combinations multiplied by the number of ways we can arrange three letters amongst themselves gives us the number of permutations.
That is,

$$3!\,_4C_3 = \,_4P_3$$

or,

$$_4C_3 = \frac{_4P_3}{3!} = \frac{4!}{(4-3)!3!} = 4$$

where $_4C_3$ is the number of combinations of four things, three at a time. Generally, the number of combinations of n things, r at a time is

$$_nC_r = \frac{_nP_r}{r!} = \frac{n!}{(n-r)!r!}$$

Example
Evaluate $_{10}C_3$.

$$_{10}C_3 = \frac{10!}{(10-3)!3!} = \frac{10!}{7!3!} = \frac{8.9.10}{2.3} = 120.$$

Example
If $_nP_r = 144$, and $_nC_r = 6$, find r.

$$r!\,_nC_r = \,_nP_r$$
$$r!.6 = 144$$
$$r! = 24 = 1.2.3.4$$
$$r = 4.$$

Example
A boy has eight friends. In how many ways can he invite three to tea? We require the number of combinations that can be formed from eight friends selected three at a time, i.e.

$$_8C_3 = \frac{8!}{(8-3)!3!} = \frac{8!}{5!3!} = \frac{6.7.8}{1.2.3} = 56$$

Example
In how many ways may nine articles be divided into two groups containing five and four articles?
The number of ways of selecting the first group is $_9C_5$. If the first group is selected, the remaining articles form the second group.

$$\text{Number of ways} = \,_9C_5 = \frac{9!}{4!5!} = 126.$$

Example
In how many ways may twelve articles be divided into groups of 6, 4, and 2, respectively?

The first group can be selected in $_{12}C_6$ ways. Out of the remaining 6 articles four may be selected in $_6C_4$ ways. The remaining two articles form the third group.

Since each of the second group can be associated with each of the first group, the number of groups is

$$_{12}C_6 \times {}_6C_4 = \frac{12!}{6!6!} \times \frac{6!}{2!4!} = 13\,860.$$

Exercise 19.5

1. Calculate $_{10}C_8$, $_{13}C_9$, $_{14}C_{12}$.
2. Prove by calculation $_{10}C_5 + {}_{10}C_4 = {}_{11}C_5$.
3. In how many ways can 30 boys be divided into two groups so that the first contains 22, and the second 8?
4. In how many ways can 60 books be divided into two groups of 40 and 20?
5. In how many ways can 60 different books be divided into three groups of 40, 15, and 5?
6. From a mixed form of 17 boys and 14 girls, 2 boys and 3 girls have to be selected. In how many ways can this be done?
7. There are 10 points in a plane, and no three are in a straight line. If the points are joined how many straight lines are formed, and how many triangles?
8. In a school a different combination of three prefects is appointed daily for playground duty. If there are 10 prefects how many weeks will elapse before the same 3 prefects are on duty together again? There are five school days in a week.
9. Show that a polygon of 10 sides has 35 diagonals.

Miscellaneous Exercise 19.6

1. What is the chance of drawing four aces from a pack of 52 playing cards?
2. A guard of 10 soldiers is to be selected from 100 soldiers. What is the probability that two particular soldiers are included?
3. Ten boys are chosen out of sixteen. What is the probability that two particular boys are included?
4. A party of six is to be chosen from ten persons. What is the probability that a particular person will be selected?
5. In the permutations of the letters of the word *permutation*, what is the probability of the two *t*'s coming together?
6. Five boys are seated on a bench. What is the probability that two particular boys will be seated at either end?
7. A committee of 3 girls and 4 boys is to be chosen from 10 girls and 17 boys. What is the probability that a particular girl and a

particular boy are on the same committee?

8. A football team of 11 players is to be chosen from 14 boys who can play equally well in any position. What are the chances of two particular boys being in the team?

9. Two balls are to be drawn from a bag containing 5 white and 7 black balls; find the chance that they will both be black.

10. If in the previous question two drawings are made, the balls being replaced before the second drawing, what is the chance that the first drawing will give two white, and the second two black balls?

   What is the chance if the first drawing consisted of two white balls and they were not replaced after the first drawing?

11. A boy has a collection of 30 records of which 20 are of classical music and 10 of pop music. In how many ways is it possible to select a group of 10 records if
   (a) no restriction is imposed on the choice?
   (b) the group is to contain 6 classical and 4 pop items?
   (c) the group is to contain 3 particular classical items?

12. A football team is to be chosen of 11 men from 16 men, of whom 5 can only play half-back, two only in goal and the remaining 9 in any of the other positions. If the team is to include only one goal-keeper and three half-backs, in how many ways can the team be chosen?

13. How many odd numbers greater than 5000 can be formed from the figures 8, 6, 5, 3, 0, if each of the figures may be used once only? What is the chance that a number will be greater than 6000?

14. In a school drama club there are 12 boys and 14 girls, and a play necessitates parts for 3 boys and 4 girls. In how many ways may the parts be allotted?

15. One boy has 10 stamps to exchange and another boy 7. In how many ways can they exchange 3 stamps?

16. Find the number of distinct ways in which the letters of the word *freeze* can be arranged. Find also the number of ways if the *e*'s are to occur alternately with the other letters. (*L*)

17. A pair of parallel lines cut a set of five parallel lines. How many different triangles can be formed from the ten points of intersection? (*L*)

18. A cricket team of 11 men is to be chosen from 15 men, of whom four can bowl and two others can keep wicket. If the team is to include only one wicket-keeper and at least three bowlers, in how many ways can the choice be made? (*L*)

19. How many odd numbers greater than 6000 can be formed from the figures 8, 6, 5, 3, 0 if each figure may be used once only? (*L*)

20. It is desired to allot the seven parts in a school play to the members of a class consisting of 12 boys and 14 girls. In how many different ways may this be done if three of the parts are to be allotted to the boys and four to the girls? Explain your method fully and leave your answer in factor form.   (L)

21. How many different numbers greater than 5000 can be formed from the digits 1, 3, 4, 5, 7 if no digit is to occur more than once in any number? Explain your method clearly.   (L)

22. The eleven members of a cricket team have to be placed in batting order. One of the players must be placed in one of the last two positions and for the first two positions only three of the players are eligible. How many different batting orders can be formed? Leave your answer in factors.   (L)

23. A housewife makes out 14 different dinner menus of which 8 have meat and 6 have fish for the main dish. If she does not use the same menu more than once, and if Sunday must be a 'meat' day and Tuesday and Friday must be 'fish' days, find the number of ways in which she can arrange dinners for a week. (Your reasons must be clearly stated.)   (L)

24. A bag contains 2 white and 3 red cubes, all of different sizes. In how many ways can 3 cubes be selected from the bag if (a) at least one cube is white, (b) at least one cube is red?   (L)

25. In how many ways can a team of 3 ladies and 3 gentlemen be selected from 10 ladies and 6 gentlemen? How many of these teams include one particular gentleman?   (L)

26. From the digits 2, 3, 4, 5, 6 how many numbers greater than 4000 can be formed? No digit may be used more than once in any number.   (L)

27. Certain registration numbers are formed from three different letters, chosen from the first ten letters of the alphabet, followed by a three-figure number which must not begin with zero. Calculate how many registration numbers can be formed.

                    (J.M.B.)

28. (a) Calculate n if (i) $_nC_3 = {}_nC_5$,
         (ii) $_nC_2 = 15$.
    (b) A committee of 8 is chosen from 10 men and 7 women so as to contain at least 4 men and at least 3 women. In how many ways can this be done?

        If one man and one woman refuse to serve together on the same committee, in how many ways can the committee be selected?   (A.E.B.)

20 The Binomial Distribution

The Binomial Expansion

A term made up of two quantities is known as a binomial term; thus, $x+y$; $a+b$; $2x-3$; $1+\frac{5}{8}x$; are all binomial terms.

The proof of the general expansion of such terms is beyond the scope of this book, but the form of the result for a positive integral index can be deduced from the consideration of particular cases.

We have by ordinary multiplication:

$(a+b)^1 = a+b$

$(a+b)^2 = a^2+2ab+b^2$

$(a+b)^3 = a^3+3a^2b+3ab^2+b^3$

$(a+b)^4 = a^4+4a^3b+6a^2b^2+4ab^3+b^4$

$(a+b)^5 = a^5+5a^4b+10a^3b^2+10a^2b^3+5ab^4+b^5.$

If the coefficients of the terms of the expansions are arranged as below, a pattern emerges—each coefficient is formed by adding the two above which lie on each side of it, as is illustrated in the triangles.

```
(a+b)¹                  1     1
(a+b)²              1     2     1
(a+b)³          1     3     3     1
(a+b)⁴       1     4     6     4     1
(a+b)⁵    1     5    10    10     5    1
```

Thus, 3 is the sum of 2 and 1; 1 is the sum of 1 and 0; 10 is the sum of 6 and 4; and so on.

This gives us a quick and easy way of expanding $(a+b)$ to any power of the index. The arrangement is known as Pascal's Triangle, and is given below up to $(a+b)^{10}$.

```
(a+b)¹                               1       1
(a+b)²                           1       2       1
(a+b)³                       1       3       3       1
(a+b)⁴                   1       4       6       4       1
(a+b)⁵               1       5      10      10       5       1
(a+b)⁶           1       6      15      20      15       6       1
(a+b)⁷       1       7      21      35      35      21       7       1
(a+b)⁸   1       8      28      56      70      56      28       8       1
(a+b)⁹  1   9     36     84    126     126     84     36      9      1
(a+b)¹⁰ 1  10    45    120    210     252    210    120     45     10     1
```

Example

Expand $(2-x)^7$.

By Pascal's Triangle,

$(a+b)^7 = a^7+7a^6b+21a^5b^2+35a^4b^3+35a^3b^4+21a^2b^5+7ab^6+b^7$.

If we put $a=2$, $b=-x$, in this expansion, we have

$(2-x)^7 = (2)^7+7(2)^6(-x)+21(2)^5(-x)^2+35(2)^4(-x)^3+35(2)^3(-x)^4$
$$+21(2)^2(-x)^5+7(2)(-x)^6+(-x)^7$$
$$= 128-488x+672x^2-560x^3+280x^4-84x^5+16x^6-x^7.$$

Example

Expand $(3x-2y)^5$.

By Pascal's Triangle,

$$(a+b)^5 = a^5+5a^4b+10a^3b^2+10a^2b^3+5ab^4+b^5.$$

Let $a=3x$; $b=-2y$; then

$(3x-2y)^5 = (3x)^5+5(3x)^4(-2y)+10(3x)^3(-2y)^2+10(3x)^2(-2y)^3$
$$+5(3x)(-2y)^4+(-2y)^5$$
$$= 243x^5-810x^4y+1080x^3y^2-720x^2y^3+240xy^4-32y^5.$$

Exercise 20.1

With the aid of Pascal's Triangle expand

1. $(1+x)^6$
2. $(1+2x)^3$
3. $(2x-3y)^4$
4. $(p+q)^8$
5. $(2x^2-3)^5$

Example

Find the value of $(1·02)^{10}$ to four decimal places.
$$(1·02)^{10} = (1+0·02)^{10}.$$

$(a+b)^{10} = a^{10}+10a^9b+45a^8b^2+120a^7b^3+210a^6b^4+$ and so on.

Let $a=1$; $b=0·02$, then,

$(1·02)^{10} = (1+0·02)^{10} = (1)^{10}+10(1)^9(0·02)+45(1)^8(0·02)^2$
$$+120(1)^7(0·02)^3+210(1)^6(0·02)^4 \text{ and so on.}$$

(There is no need to expand further as the omitted terms do not affect the fourth decimal place.)

$$= 1+0·2+0·0180+0·00096+0·0000672$$
$$= 1·219027$$
$$= 1·2190.$$

Example

Show that if x^3 and higher powers of x can be neglected
$$(1+x)^2(1-2x)^8 = 1-14x+81x^2$$

$(1+x)^2 = 1+2x+x^2$

$(1-2x)^8 = 1-16x+112x^2$ (since x^3 and higher powers are neglected)

$(1+x)^2(1-2x)^8 = (1+2x+x^2)(1-16x+112x^2)$
$$= 1-16x+112x^2+2x-32x^2+x^2$$
$$= 1-14x+81x^2.$$

Exercise 20.2

1. Expand (a) $(x+y)^4$; (b) $(1-x)^5$; (c) $(x-y)^4$; (d) $(x+2y)^5$; (e) $(x+2)^7$; (f) $(3x-2y)^8$.

2. Find (a) the fourth term of the expansion of $(p+q)^9$
   (b) the third term of the expansion of $(1-x)^{12}$
   (c) the fifth term of the expansion of $(x+y)^{11}$
   (d) the sixth term of the expansion of $(m+n)^{10}$
   (e) the seventh term of the expansion of $(a+b)^{13}$

3. Find correct to four decimal places the value of
   (a) $(1\cdot02)^5$; (b) $(1\cdot03)^9$; (c) $(1\cdot2)^6$; (d) $(3\cdot01)^5$; (e) $(1\cdot98)^4$.

4. Show that if x^2 and higher powers of x can be neglected
   $$(1-x)^3(1+3x)^5 = 1+12x.$$

5. If $p = \frac{1}{3}$; $q = \frac{2}{3}$, find the value of the following terms in the expansion of $(p+q)^{11}$.
   (a) 3rd. (b) 5th. (c) 8th. (d) 10th. (e) 4th.

6. (a) Find the middle term in the expansion of $(2x+y)^6$.
   (b) Find the term independent of x in the expansion of $\left(x^2+\dfrac{1}{x}\right)^6$.

   (L)

7. (a) Expand $(2x-\frac{3}{4})^4$, writing each term in its simplest form.
   (b) Find the value of the term independent of x in the expansion of $\left(x-\dfrac{1}{x^2}\right)^6$.

   (L)

8. (a) Find the coefficient of x^2 in the expansion of $\left(x-\dfrac{2}{x^2}\right)^8$.
   (b) Find the coefficient of x^3 in the expansion of the product
   $$(1+x)^6(1-3x).$$

   (L)

9. (a) Write down the fifth term in the expansion of $(x+\frac{1}{2})^{10}$ in descending powers of x and find its value when $x = 2$.
   (b) Find the middle term in the expansion of $\left(2x-\dfrac{1}{3x}\right)^6$. (L)

10. Show that, if x^3 and higher powers of x can be neglected,
    $$(1-x)^2(1+3x)^6 = 1+16x+10x^2.$$ (A.E.B.)

The Binomial Distribution

The second theorem on probability states that, if we have two independent events, then the probability of both happening is the product of the probabilities of each happening.

Let us now determine what happens in the case, say, of a number of dice thrown together, and find the probabilities of the appearance of so many sixes at each throw. Let the appearance of a six be considered a success, and the appearance of any other number a failure. Let s stand for success, and f for failure, and let p be the probability of success, and q the probability of failure. In this particular example $p = \frac{1}{6}$, and $q = \frac{5}{6}$.

For one die there are two possible events, namely s or f.

For two dice there are four possible events, because we can combine the s or f of the first die with the s or f of the second die, and may express the four results as

$$ss, \; sf, \; fs, \; ff,$$

the meaning of which is, we may get a six followed by a six, a six followed by some other number, some other number followed by a six, some number other than a six followed by some number other than a six.

For three dice there are eight possibilities, because we can combine the s or f of the third die with each of the four previous events, thus getting

$$sss, \; sfs, \; fss, \; ffs, \; ssf, \; sff, \; fsf, \; fff,$$

and so on for four or more dice.

If the order of success or failure is unimportant, we may consider terms like ssf, sfs, fss, as being equivalent, and the above results can then be written more concisely as follows:

One die			s	f		
Two dice			ss	$2sf$	ff	
Three dice		sss	$3ssf$	$3sff$	fff	
Four dice		$ssss$	$4sssf$	$6ssff$	$4sfff$	$ffff$
Five dice	$sssss$	$5ssssf$	$10sssff$	$10ssfff$	$5sffff$	$fffff$

and so on.

Now the probability of an s is p, and of an f is q. Therefore, by the second theorem on probability, the probability of each event in the above table is

One die			p	q		
Two dice			p^2	$2pq$	q^2	
Three dice		p^3	$3p^2q$	$3pq^2$	q^3	
Four dice		p^4	$4p^3q$	$6p^2q^2$	$4pq^3$	q^4
Five dice	p^5	$5p^4q$	$10p^3q^2$	$10p^2q^3$	$5pq^4$	q^5

This means that if, say, three dice are cast, the probability of obtaining three successes, that is three sixes, is p^3, or $(\frac{1}{6})^3$; of obtaining two successes and one failure, $3p^2q$, that is $3(\frac{1}{6})^2 . (\frac{5}{6})$; of obtaining one success and two failures, $3pq^2$, that is $3(\frac{1}{6}).(\frac{5}{6})^2$; that of obtaining three failures, q^3, that is $(\frac{5}{6})^3$.

On examining the table, we see the probabilities for the different events are, in the case of one die, the terms of the expansion $(p+q)^1$; in the case of two dice, the terms of the expansion of $(p+q)^2$; in the case of three dice, the terms of the expansion of $(p+q)^3$; and so on. Therefore we conclude that, if n dice are thrown, the probabilities for each possible event will be the terms of the expansion of $(p+q)^n$.

Example

If ten dice are cast, what is the probability of obtaining exactly seven sixes?

By the binomial expansion
$$(p+q)^{10} = p^{10} + 10p^9q + 45p^8q^2 + 120p^7q^3 + 210p^6q^4 + 252p^5q^5$$
$$+ 210p^4q^6 + 120p^3q^7 + 45p^2q^8 + 10pq^9 + q^{10}.$$

Therefore the probability of obtaining exactly seven sixes is $120p^7q^3$, where $p = \frac{1}{6}$, and $q = \frac{5}{6}$, that is $120(\frac{1}{6})^7.(\frac{5}{6})^3 = 0.00028$.

Similarly the chance of obtaining exactly
$$\text{five sixes is } 252p^5q^5 = 0.013,$$
$$\text{and two sixes } 45p^2q^8 = 0.29,$$
$$\text{and ten sixes } \quad p^{10} = 1.65 \times 10^{-8}.$$

Probability Trees

A useful device for displaying the probabilities discussed in this chapter is a *probability tree*. Suppose we take the case of 4 dice to demonstrate the properties and construction of these diagrams; a success again being represented by $p = \frac{1}{6}$ and a failure represented by $q = \frac{5}{6}$. The diagrams show the outcomes for 1, 2, 3 and 4 dice.

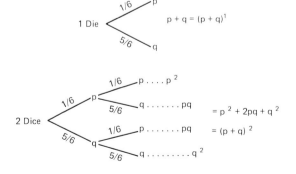

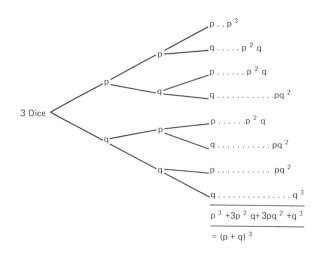

3 Dice

$p \ldots p^3$
$q \ldots p^2 q$
$p \ldots p^2 q$
$q \ldots pq^2$
$p \ldots p^2 q$
$q \ldots pq^2$
$p \ldots pq^2$
$q \ldots q^3$

$$p^3 + 3p^2 q + 3pq^2 + q^3$$
$$= (p + q)^3$$

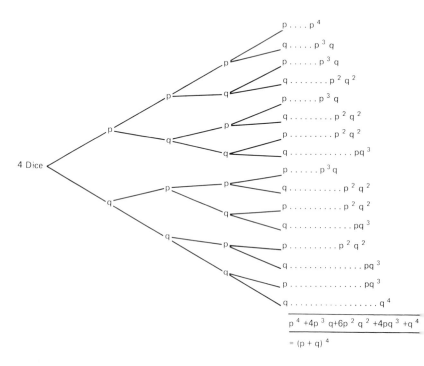

4 Dice

$p \ldots p^4$
$q \ldots p^3 q$
$p \ldots p^3 q$
$q \ldots p^2 q^2$
$p \ldots p^3 q$
$q \ldots p^2 q^2$
$p \ldots p^2 q^2$
$q \ldots pq^3$
$p \ldots p^3 q$
$q \ldots p^2 q^2$
$p \ldots p^2 q^2$
$q \ldots pq^3$
$p \ldots p^2 q^2$
$q \ldots pq^3$
$p \ldots pq^3$
$q \ldots q^4$

$$p^4 + 4p^3 q + 6p^2 q^2 + 4pq^3 + q^4$$
$$= (p + q)^4$$

Exercise 20.3

1. Calculate the probability of obtaining one six when three dice are thrown.

2. Calculate the probability of obtaining exactly four heads when seven coins are tossed.

3. The probability of a marksman hitting the bull is $\frac{3}{4}$. What are his chances of hitting the bull three or more times in six shots?

4. A card is drawn from each of four different packs. What is the probability of drawing exactly 4, 3, 2, 1, or 0 aces?

5. In twelve throws of a coin what is the chance of obtaining ten or more heads?

6. Five red and ten yellow otherwise identical balls are placed in a bag. If four balls are selected what are the chances of getting either 0, 1, 2, 3, or 4 red balls? (Each time the selected ball is replaced and the balls thoroughly mixed).

7. Two men, **A** and **B**, play a game. **A**'s chance of winning is $\frac{3}{5}$. What is **A**'s chance of winning 4 or more games out of a series of 8 games?

8. During a cold epidemic in a school, the chance of catching a cold is 20%. What is the chance that, out of five boys, three or more will catch cold?

9. Five dice are thrown a large number of times. The occurrence of an ace is considered a success. In what percentage of the trials would exactly four successes be expected?

The Mean and Standard Deviation of a Binomial Distribution

The probability of obtaining a head on the toss of a coin is $\frac{1}{2}$. Therefore in, say, ten trials we should expect heads $10 \times \frac{1}{2}$, or five times. Similarly, if the probability of an event is $6p^5q^6$, say, then in k trials we should expect the event to occur $k6p^5q^6$ times.

Now, if four coins are tossed together, the probability of a head occurring a certain number of times is given by the terms of the expansion of $(p+q)^4$, where $p = \frac{1}{2}$ and $q = \frac{1}{2}$.

$$(p+q)^4 = p^4 + 4p^3q + 6p^2q^2 + 4pq^3 + q^4$$

Therefore in, say, 160 tosses of four coins, the frequency with which a certain number of heads would be expected to occur is as follows:

$$\text{four times is } 160(\tfrac{1}{2})^4 \qquad = 10$$
$$\text{thrice} \qquad 160.4.(\tfrac{1}{2})^3.(\tfrac{1}{2}) \quad = 40$$
$$\text{twice} \qquad 160.6.(\tfrac{1}{2})^2.(\tfrac{1}{2})^2 = 60$$
$$\text{once} \qquad 160.4.(\tfrac{1}{2}).(\tfrac{1}{2})^3 \quad = 40$$
$$\text{not at all} \qquad 160.(\tfrac{1}{2})^4 \qquad = 10.$$

This is the frequency distribution of the number of times heads is expected to occur, and we can calculate its mean and standard deviation.

Number of heads appearing	Frequency f	Deviation from Working Mean, 0 d	fd	fd^2
0	10	0	0	0
1	40	1	40	40
2	60	2	120	240
3	40	3	120	360
4	10	4	40	160
Totals	160		320	800

Arithmetic mean = working mean + average of deviations from working mean

$$= 0 + \frac{320}{160}$$
$$= 2$$

$$\text{and,} \quad s^2 = \frac{800}{160}$$
$$= 5$$

$$\text{and} \quad \sigma^2 = s^2 - b^2$$
$$= 5 - 4 \quad (b = 2)$$
$$= 1.$$

Now suppose four similar but independent events occur, (like the tossing of four coins), and the probability of success in each event is p, and of failure q. Then in k trials the number of times success is expected

$$\text{four times is } kp^4$$
$$\text{thrice} \qquad \text{is } k4p^3q$$
$$\text{twice} \qquad \text{is } k6p^2q^2$$
$$\text{once} \qquad \text{is } k4pq^3$$
$$\text{not at all} \quad \text{is } kq^4$$

This is a frequency distribution and we can find its mean and standard deviation. Remember $p + q = 1$.

Number of times success occurs	Frequency f	Deviation from Working Mean, 0 d	fd	fd^2
0	kq^4	0	0	0
1	$k4pq^3$	1	$k4pq^3$	$k4pq^3$
2	$k6p^2q^2$	2	$k12p^2q^2$	$k24p^2q^2$
3	$k4p^3q$	3	$k12p^3q$	$k36p^3q$
4	kp^4	4	$k4p^4$	$k16p^4$
Total	k		$4kp$	$4kp(1+3p)$

Method:

Total $f = k(q^4 + 4pq^3 + 6p^2q^2 + 4p^3q + p^4) = k(q+p)^4 = k.$

Total $fd = k4p(q^3 + 3pq^2 + 3p^2q + p^3) = k4p(q+p)^3 = 4kp.$

Total $fd^2 = k4p(q^3 + 6pq^2 + 9p^2q + 4p^3)$

$\qquad = k4p[(q^3 + 3pq^2 + 3p^2q + p^3) + 3p(q^2 + 2pq + p^2)]$

$\qquad = k4p[(q+p)^3 + 3p(q+p)^2]$

$\qquad = k4p(1 + 3p).$

Arithmetic mean = working mean + average of deviations from working mean

$$= 0 + \frac{4kp}{k}$$

$$= 4p.$$

and, $\sigma^2 = 4p(1+3p) - 16p^2$

$\qquad = 4p(1 + 3p - 4p)$

$\qquad = 4p(1 - p)$

$\qquad = 4pq$

Therefore $\sigma = \sqrt{4pq}.$

Now calculate the mean and standard deviation for the frequency distributions of the number of successes with *two* and *three* independent events occurring k times. The answers are: means $2p$ and $3p$, and the standard deviations $\sqrt{2pq}$ and $\sqrt{3pq}$. Therefore we conclude that if n events occur a number of times, the mean number of times success occurs is np, and the standard deviation is $\sqrt{npq}$.

These two results are very important.

The mean of a binomial distribution is np, and its standard deviation is $\sqrt{npq}$.

Meaning of the Results

It is essential to understand what these results measure, and the following examples may help to make it clear.

Example

If we toss 400 coins, counting 'head' a success, the number of heads we should expect to get is the mean np, i.e., $400 \times \frac{1}{2}$ or 200. The second result gives us a measure of the unlikelihood of obtaining the expected mean number of successes. The greater the measure of deviation, the greater the chance of missing the mean, and it is not unlikely that the number of successes obtained may differ from the mean by as much as one standard deviation or $\sqrt{npq}$, i.e. by $\sqrt{400 \times \frac{1}{2} \times \frac{1}{2}}$, or 10.

Therefore the probability is 0·68 that the number of heads obtained in any one experiment will lie between 200 ± 10, i.e., 190 and 210. (See page 193.)

Example

If, say, 200 dice are thrown and we count obtaining a six a success, the probability of obtaining a success with one die is $\frac{1}{6}$, and a failure $\frac{5}{6}$.

Therefore the mean number of successes expected is $200 \times \frac{1}{6} = 33·3$, and the expected deviation is $\sqrt{200 \times \frac{1}{6} \times \frac{5}{6}} = 5·3$, i.e. the chances are 0·68 we should obtain between $33·3 \pm 5·3$, or, 28 to 39 sixes.

At times it is more convenient to refer to the proportion of successes expected in any experiment. The proportion of successes expected is $\dfrac{np}{n}$, i.e. p, and it is unlikely that the proportion will differ from p by more than $\dfrac{\sqrt{npq}}{n}$, i.e., $\sqrt{\dfrac{pq}{n}}$.

Exercise 20.4

1. Find the mean and standard deviation of a binomial distribution in which (*a*) $n = 2$, $p = \frac{1}{2}$; (*b*) $n = 3$, $p = \frac{1}{2}$; (*c*) $n = 5$, $p = \frac{1}{3}$.
2. The values of the mean and standard deviation of a binomial distribution are 12 and 3, respectively. Calculate n, p, and q.
3. One third of the pupils in each form in a school take school dinner. The school is large and has 100 forms. If a sample of five pupils is taken from each form, in how many samples would it be expected to find exactly three pupils taking school dinner?
4. Assuming that three footballers in every five smoke, what is the probability of finding (*a*) a soccer team, (*b*) a rugby team composed of non-smokers?

    How many smokers would you expect to find in (*c*) a soccer team, (*d*) a rugger team?

5. From each of four packs of playing cards select a card at random. The selection of a club is to be considered a success. Repeat 200 times, each time replacing the selected card and shuffling each pack. List the number of times 4, 3, 2, 1, or 0 clubs are selected, and compare the results with the theoretical frequencies.

6. If a six-figure number is formed by selecting at random six figures from 0, 1, 2, 3, 4, 5, 6, 7, 8, 9, where each figure can be selected any number of times, calculate the probability of, say, 0 appearing not at all, once, twice,...six times. From a list of Premium Bond winning numbers, count the number of times 0 appears not at all, once, twice,...six times in each number. Compare the result with the theoretical estimate, and the mean and standard deviation with the theoretical values.

7. What is the chance of correctly forecasting the results of 10 or more matches out of 12 matches?

8. Toss four coins 200 times. Tabulate the results to show the frequency of heads turning up 4, 3, 2, 1, 0 times. Compare the result with the theoretical result.

9. If the probability of a new-born child being a boy is 52%, what is the probability that a family of four children consists of three boys and a girl?

10. A manufacturer finds that one article in every twenty is below the required standard. How many substandard articles would he expect to find in a sample of 200?

11. A pupil is set a test of twelve questions, each requiring an answer of 'yes' or 'no'. If the pupil knows nothing about the subject, and makes a guess at each answer, what is the probability of his passing if the pass mark is seven or more correct answers?

12. On a normal school day the chances of a pupil being absent from school is $\frac{1}{20}$. What is the probable number of pupils absent on a normal day in a school of 800 pupils?

13. The average number of frosty days at Kew Observatory in December between 1914 and 1940 was 9. Taking this as the probability of any particular day in December being frosty, what are the chances that there are six or more frosty days in any week in December?

14. An electric bulb manufacturer finds that 4% of the bulbs are defective. How many perfect bulbs are to be expected in a sample of 400?

   What are the chances that a random sample of 20 will not contain more than one defective bulb?

15. The probability of a footballer scoring a goal in any game is $\frac{1}{3}$. How many goals would you expect him to score in thirty games?

16. A manufacturer of stockings sells them packed six pairs in a box. On testing 100 boxes for faulty pairs of stockings he obtained the following results.

Number of faulty pairs	0	1	2	3	4	5	6
Number of boxes	82	12	4	2	0	0	0

Calculate the mean number of faulty pairs of stockings per box.

Assuming that it is a binomial distribution calculate the probability that a random pair of stockings will be faulty.

17. The mean of a binomial distribution is 20, and the standard deviation is 4. Calculate n, p, and q.

18. Assuming that the number of aliens registered in the United Kingdom is one in three hundred of the population, and that a football crowd is a random sample of the population, how many aliens would you expect to find in a crowd of 60 000?

19. The chances that a man aged 30 years dies before the age of 60 years is $\frac{1}{4}$. What is the probability that of a group of 12 men aged 30 years at least 8 will attend a reunion in 30 years' time?

20. Two theoretical binomial distributions have the same total frequency, and the ratio of the variance of the first to the variance of the second is $\frac{32}{27}$. If the probability of success in the first distribution is $\frac{1}{3}$, what is the probability of success in the second distribution?

21. In two binomial distributions $\frac{n_1}{n_2} = \frac{1}{3}$, $\frac{\text{mean}_1}{\text{mean}_2} = \frac{5}{9}$, and $\left[\frac{\sigma_1}{\sigma_2}\right]^2 = \frac{100}{189}$. Calculate p_1 and p_2.

22. Explain what is meant by a 'binomial distribution'.

A hen may be expected to lay five eggs a week. If six hens are kept, calculate the chance of getting (a) six eggs in one day, (b) four eggs in one day. (C.A.)

23. (a) In a binomial distribution where the frequencies of occurrence of an event are given by the terms of the expansion $N(p+q)^m$, the mean μ and the standard deviation σ of the distribution are given by the formulae
$$\mu = mp, \quad \sigma = \sqrt{mpq}.$$
Prove these formulae in the case where $m = 3$.

(b) Two men **A** and **B** play a game in which **A** should win 8 games to every 7 won by **B**. If they play 3 games, show that the probability that **A** will win at least two games is approximately 0·55. (C.A.)

24. A battery of four guns is firing on to an enemy emplacement. It is reckoned that each gun should score on the average one direct hit in every five shots, and that three direct hits are needed to destroy the emplacement. If each gun fires one shell, calculate the probability that the emplacement will be destroyed.

    With new gun crews it is reckoned that two of the guns should score one direct hit in every three shots and that the other two guns should score one direct hit every four shots. If each gun now fires one shell, calculate the probability that the emplacement will be destroyed. (*CA.*)

25. If hens of a certain breed lay eggs on four days a week, on the average, find how many days during a season of 200 days a poultry keeper with eight hens of this breed will expect to receive at least six eggs. (*CA.*)

26. If 5 coins are tossed a large number of times, the frequencies with which 0, 1, 2, 3, 4, 5 heads occur may be taken as proportional to the coefficients of $(1 + x)^5$.

    Write down an expected frequency distribution if the coins are tossed 2880 times. On how many occasions would you expect to obtain 2 or more heads? (*L*)

27. When three pennies are tossed simultaneously a large number of times, it may be assumed that the frequencies with which 0, 1, 2 or 3 heads occur are proportional to the coefficients of $(1 + x)^3$. Write down an expected frequency distribution when the three pennies are tossed simultaneously 16488 times. (*L*)

28. Four coins are tossed simultaneously. Find the probability that
    (i) all four coins come down 'heads',
    (ii) two and only two come down 'heads',
    (iii) two or more than two come down 'heads'. (*L*)

29. Assuming the chances of a baby being a boy or being a girl to be equal, find
    (i) the probability that a family of six children consists of three boys and three girls,
    (ii) the probability that a family of six children contains at least four girls. (*L*)

30. In two binomial distributions the ratios of the number of trials and of the arithmetic means are 64:75 and 16:15, respectively, and the variances are equal. Calculate the probability of a 'success' in the first and second distributions. (*A.E.B.*)

Miscellaneous Exercises

1. The following table shows the number of vehicles passing 117 points in a large city during a day in 1977 and the percentage increase in vehicles passing the same points during a day in 1979.

Time Period	Number of Vehicles (1977)	Percentage Increase (1979 compared with 1977)
	(000's)	
8 to 9.59 a.m.	658	10·8
10 to 11.59 a.m.	561	3·3
12 noon to 1.59 p.m.	511	5·7
2 to 3.59 p.m.	545	4·4
4 to 5.59 p.m.	706	9·8
6 to 7.59 p.m.	573	10·8

Calculate:
   (a) the average percentage increase in traffic over the whole of the 12-hour period;
   (b) the volume of traffic passing during the midday period (10 a.m. to 3.59 p.m.) expressed as a percentage of the traffic during the whole 12-hour period in each of the two years.   (*I.T.*)

2. 12 500 people showed average daily earnings of £10.8. (a) Calculate the total earnings if the number of people is taken to the nearest hundred and the earnings to the nearest £0.1. (b) Calculate the relative error in the total earnings. (c) Explain the difference between relative errors and unbiased errors and state the circumstances under which relative errors would be of the greatest use.
    (*R.S.A.I.*)

3. $S = P + Q$; S is estimated at 156 000 correct to the nearest 1000, and P at 49 500 correct to the nearest 100. Find the value of $\dfrac{S}{Q}$ with the possible limits of error.   (*I.S.I.I.*)

4. From the following data, calculate the maximum and minimum populations consistent with the figures in the table, remembering that the 'per 1000' entries are computed to the nearest unit in the first decimal place.

England and Wales 1975

	Number	Per 1000 of Population
Births	603 445	12·3
Deaths	582 841	11·8
Marriages	380 620	15·5

5. The following figures record the number of motor vehicles passing a road junction during a day in 1960, and the percentages by which traffic increased between 1960 and 1964.

Time	Number of vehicles in 1960	Percentage increase between 1960 and 1964
08.00–09.59	7320	9·9
10.00–11.59	6120	2·4
12.00–13.59	6020	4·8
14.00–15.59	5090	3·5
16.00–17.59	7410	8.9
18.00–19.59	6150	9·5

Calculate:
(a) The average percentage increase in traffic over the whole of the 12-hour period;
(b) For each year, the number of vehicles passing during the evening period (4 p.m.–7.59 p.m.) as a percentage of the whole 12 hours traffic. (*I.T.*)

6. In one day (12th May, 1978) the output of shoes from a certain factory was as follows. Boy's and men's brown shoes under size 3, 305 pairs; from size 3 to under size 5, 355 pairs; from size 5 to size 7, 410 pairs; over size 7, 240 pairs. The corresponding figures for black shoes were 225, 300, 350 and 210. In addition 260 pairs of plastic sandals were made. Women's shoes were also made in the same sizes, except that none was made over size 7. The figures for brown were 300, 450 and 150; and for black 255, 410 and 110. On the following three days the total output each day was; for men's shoes 2524 pairs, 2590 pairs and 2622 pairs; and for women's shoes 1760 pairs, 1890 pairs and 2065 pairs.

Tabulate this data in a suitable form with a title.

7. Between 1969 and 1977 the number of people found guilty of indictable offences in England and Wales rose by 58·7% to a total

of 428 700. The number found guilty of theft and handling stolen goods rose by 49 800 to 236 600. The corresponding figures for burglary were 3500 and 70 400, for fraud and forgery 6500 and 21 000, for wounding 17 400 and 37 000, and for criminal damage 40 700 and 42 100. However, those convicted of social offences fell by 300 to 6200.

Tabulate this data, calculating any necessary figures.

8. In the period 1975 to 1977 the number of pupils in schools rose by 46 504 to 9 663 978. The number of those aged between 2 and 4 years fell by 21 884 to 319 400 and there was also a fall in the number of those aged between 5 and 14 years of 32 068 to 795 657. The increased number came from pupils aged 15 years and over. During the same period the number of teachers rose by 14 526 to 506 829 and the number of schools rose by 44 to 33 129.

Tabulate this data in a suitable form.

9. Consider the main problems of classification in statistical work. Illustrate your answer by reference to any official statistics with which you are familiar. (*L.G.I.*)

10. Road Goods Vehicles: Analysis by Unladen Weight
Great Britain

	Thousands	
	1966	1976
not over 1 ton	529	600
1 ton not over 2 tons	391	572
2 tons not over 3 tons	176	132
3 tons not over 5 tons	271	152
5 tons not over 8 tons	88	127
8 tons	29	102

SOURCE: Dept. of Transport

Draw a bar chart which illustrates the above data and compares the differences between the figures for each of the years.

11. Twelve thousand glass bulbs are supplied to make 40-watt, 60-watt, and 100-watt electric-light bulbs. Bulbs are rejected at several manufacturing stages for different faults.

At stage 1, 10 per cent of the 40-watt, 4 per cent of the 60-watt, and 5 per cent of the 100-watt bulbs are broken. At stage 2, about 1 per cent of the remainder of the lamps have broken filaments. At stage 3, 100 hundred-watt lamps have badly soldered caps and half as many have crooked caps. Twice as many 40-watt and 60-watt lamps have these faults. At stage 4 about 3 per cent are rejected for bad type-marking and 1 in every 100 is broken during the packing which follows.

Arrange this information in concise tabular form. Which type of lamp shows the greatest wastage during manufacture?

(R.S.A.A.)

12. In 1977 the number of new houses built for private owners was 114855 and this was 8547 less than the number for the previous year. Of this total for 1977, 1025 had 1 bedroom, 21663 had 2 bedrooms, 72816 had 3 bedrooms and the remainder had 4 or more bedrooms. The corresponding figures for 1976 were; 1048, 22248 and 79286. The number of flats built for private owners also fell by 1803 to 13272 and of these 65 had 4 or more bedrooms, 987 had 3 bedrooms, 8585 had 2 bedrooms and the remainder had one bedroom. The corresponding figures for 1976 were; 44, 1294 and 9793.

Draw up a table with suitable headings and totals to display this data.

13.

Consumers' Expenditure at current prices (£M)

Year	Total	On running cost of vehicles	On other travel
1969	29104	1595	941
1970	31696	1692	1019
1971	35399	1849	1161
1972	39944	2114	1298
1973	45201	2407	1473
1974	51977	3090	1648
1975	63552	3874	2077
1976	73656	4418	2467

Draw a chart to bring out the year to year changes in the amount of consumer spending devoted to vehicles and other travel, respectively. Give brief reasons for your choice of method of presentation.

14. In 1976, information derived from 'Family Expenditure' showed that the percentage of households in the United Kingdom having the following durable goods was as shown.

Washing Machine	72·3%
Refrigerator	88·1%
Television	95·6%
Telephone	52·6%

Using a radius of 5 cm, display this information in a pie chart.

15. The following table shows the lives (in hundreds of km) of eighty tyres removed from double-deck buses.

1464	1167	1067	1406	779	1120	1266	1006
1714	1434	1763	2167	1440	1041	827	1720
1012	1109	1458	1831	727	698	938	1270
1670	1286	1435	1496	166	980	1318	1656
1456	1146	1184	1099	1630	1364	1725	1207
1371	1007	1170	1563	1567	1474	1693	1047
1530	861	1155	1595	1582	1895	1619	716
1455	962	1186	1371	1027	1558	1638	455
1879	632	552	998	1006	1691	1974	1709
1424	1710	1924	1276	1012	932	1064	1303

Group the above figures into a suitable frequency distribution and represent them (*a*) by a histogram, and (*b*) by a frequency polygon.   (*I.T.*)

16. The value of the output of a small factory in £ thousand is shown for the period January to September with the figure for June missing. Plot this data using 2 cm on the horizontal axis for each month of the year and start the vertical axis at £10 000 with 1 cm equal to £2000. Put in the 'line of best fit' and extend it to the end of the year. Interpolate to find the probable figure for June and extrapolate to estimate the value for December if the same trend continues.

Month	Jan.	Feb.	Mar.	Apr.	May	July	Aug.	Sep.
Output (£ Thousand)	18·3	20·1	21·4	23·3	24·8	28·3	30·0	32·3

17. The distribution of hospitals and allocated beds in Great Britain at 30th June, 1976 is shown below. Display the information given on a suitable diagram and comment on any feature of interest.

Size of Hospital (beds)	No. of Hospitals	No. of Allocated Beds
under 50	981	26,436
50 to 249	1105	125,564
250 to 499	304	110,106
500 to 999	208	140,955
1000 to 1999	59	76,298

SOURCE: D.H.S.S.

282

18. Comment on the use of the following forms of graphical presentation illustrating your answer with sketches:
    (*a*) Pie charts
    (*b*) Bar charts
    (*c*) Correlation diagrams. (*I.S.A.*)

19. Suggest suitable types of chart or diagram to show:
    (*a*) The progress of a company during the year towards estimated annual profits.
    (*b*) Annual surplus or deficit in U.K. balance of payments over a ten-year period.
    (*c*) Registered unemployed males and females for the past ten years.
    (*d*) Age distribution of population in 1851 and 1951. (*I.S.A.*)

20. Write explanatory notes on the following:
    (*a*) Ratio scales
    (*b*) Bar charts
    (*c*) Pictorial statistics. (*R.S.A.I.*)

21. The net output per person employed (£) in the manufacturing industries for the years 1970–75 is given in the table together with the average number of persons employed in these industries during this period. Plot both sets of values on the same graph and comment on any features revealed.

Year	Net Output/person (£)	Av. No. of Persons Employed (Thousand)
1970	2397	8033
1971	2633	7830
1972	3029	7522
1973	3487	7613
1974	4206	7700
1975	4928	7504

SOURCE: B.S.O.

22.

Inland Revenue Duties (1975–6)	
Source	£m.
Income Tax	15,053·9
Surtax	108.8
Corporation Tax	1,997·3
Capital Gains Tax	387·1
Estate Duty	212·5
Capital Transfer Tax	117·7
Stamp Duties	281·1
Total	18,158·4

SOURCE: H.M. Treasury

Represent the above information graphically.

23.

Death Rates per Million
by Sex and Age

Age Group (years)	England and Wales		Greater London		Urban Areas		Rural Areas	
	Males	Females	Males	Females	Males	Females	Males	Females
0–14	105	48	65	34	100	48	130	52
15–44	173	27	133	29	174	26	267	33
45–64	134	43	117	44	123	35	183	55
65 and over	316	135	363	184	286	115	293	85

Analyse this information by means of graphs and secondary statistics and summarise your results in the form of a report.

(*R.S.A.A.*)

24.

Year	1968	1969	1970	1971	1972	1973	1974	1975	1976
Profit (£ Thousand)	2993	4088	5122	6451	6617	8370	11326	15191	23388

Plot these figures to show the rate of growth of profit.

25. 'In 1912 there were 250000 road vehicles in Britain. Twenty years later there were ten times as many. By 1952 the number had doubled again, to double yet again by 1962 when there were 10½ millions. There is now the prospect of 18 million vehicles by 1970, 25 millions by 1980 and more than 30 millions by the end of the century.' (*The Guardian*, 7th Nov. 1966)

Draw a graph to illustrate this statement and comment on the shape of the graph. (*I.S.I.*)

26.   Index of Industrial Production
                          (Base 1970 = 100)
                   1969   99·8
                   1970   100
                   1971   100·4
                   1972   102·7
                   1973   110·2
                   1974   106·3

(a) Plot the above data on a semi-logarithmic graph.

(b) Comment on the change in industrial production between 1969 and 1974 in the light of your graph.

27. What are the properties of the semi-logarithmic graph? Compare the relative merits of such graphs with other methods of representing a time series. Illustrate your answer with reference to any field of official statistics. *(L.G.I.)*

28. (a) Describe concisely the systematic or quasi-random method of sampling.

    (b) (i) A given population contains 10 000 items. Find the largest sample which can be obtained by systematic sampling every 50th item, the random selection of the first being 37.

        (ii) A sample of 100 employed persons is required from the Ministry of Labour's file of cards. Describe how you would obtain this sample. *(R.S.A.A.)*

29. What are the main considerations to be taken into account in (a) the sample design, and (b) the content of the questionnaire, of a sample survey intended to find out the reaction of various sections of the community to a plan for a by-pass round the centre of a town? *(I.S.I.)*

30. In what circumstances and in what way does stratification improve sample design? *(L.G.F.)*

31. The following figures are the arrival times of fifty long-distance trains, expressed in terms of the number of minutes late (minutes early shown as minus) compared with the scheduled arrival times.
69, 1, −8, −9, −3, 84, 37, 47, 26, −7, 68, 8, 39, −2, 4, 33, −2, 22, −3, 18, 13, 56, −2, 0, −18, 0, −2, 32, 28, −23, 10, −5, 13, −10, 4, 24, 45, 20, 1, −11, 0, 5, −3, 13, 18, 17, 21, 6, −10, 19.

Group the figures into a suitable frequency distribution and represent them (a) by a histogram and (b) by a cumulative frequency diagram. *(I.T.)*

32. What is meant by 'stratified random sampling'? Explain the procedures and advantages of stratification in sample design.

(*L.G.F.*)

33. Draft a questionnaire to be sent through the post to a sample of householders in order to obtain information regarding the number of cycles owned and the purposes for which they are used.

(*I.S.A.*)

34. A company operating 100 goods vehicles decides to change the composition of its fleet, with the following results:

	Before Change		After Change	
Size of vehicle	No. of vehicles in fleet	Average weekly distance per vehicle	No. of vehicles in fleet	Average weekly distance per vehicle
Small	25	233 km	20	203 km
Medium	45	418 km	15	318 km
Large	30	731 km	65	635 km

What is the average distance travelled per vehicle per week (*a*) before the change and (*b*) after the change? Explain the apparent contradiction between your results and the comparison between the figures in the second and fourth columns of the table. Indicate how you would set about calculating an alternative comparison.

(*I.T.*)

35. The following figures are the running times (in minutes) of a sample of buses on a congested section of route:
16, 19, 20, 17, 16, 18, 17, 16, 15, 16, 18, 16, 18, 19, 15, 15, 15, 14, 14, 17, 17, 19, 25, 16, 17, 17, 13, 18, 16, 19, 16, 17, 15, 19, 18, 15, 22, 16, 20, 16, 15, 17, 19, 18, 15, 18, 15, 15, 16, 17.
Estimate:
(*a*) the median; (*b*) the mode; (*c*) the two quartiles.   (*I.T.*)

36. Calculate the arithmetic average, the standard deviation and the coefficient of variation from the following data:
Concentrations of solution (grammes per litre):
14·2 15·6 13·7 12·9 13·4 13·6 14·0 15·1 14·5 15·0
Explain briefly the uses of the coefficient of variation. (*I.S.I.*)

37. In order to determine a scheduled running time for buses over one section of a route, the actual running times (in minutes) were noted in respect of a sample of 60 trips, as follows:

15 17 16 17 17 15 18 13 17 16 16 17 19 18 20 25 15 16
17 16 19 19 15 18 15 19 16 17 18 16 16 15 15 16 21 15
17 19 15 22 17 18 16 18 14 16 15 19 24 16 14 20 19 25
18 18 17 16 21 18

Estimate, explaining in each case your method of calculation, the arithmetic mean, the median, the mode, the upper quartile and discuss the advantages and disadvantages of each of these measures if used as a basis for fixing the scheduled running time for this section of the route. (*I.T.*)

38. Using a four-quarter centred moving average, plot the data given and put in the trend line. Calculate the average relative seasonal variation for each of the four quarters and hence find the de-seasonalised values of the data.

Value of new housing in Private Sector (£M)

Year	Quarters			
	1	2	3	4
1971	—	—	226	220
1972	250	322	306	320
1973	421	409	382	301
1974	212	232	—	—

SOURCE: Dept. of Environment

39. A survey of average speeds on the main roads of Glasgow gave the following results:

Speed (km/h.)	Per cent of traffic travelling at these speeds	
	Central Glasgow	Outer area of Glasgow
Under 10	18	—
10–12·9	35	—
13–15·9	26	2
16–18·9	13	6
19–21·9	8	17
22–24·9	—	30
25–27·9	—	25
28–30·9	—	10
31–33·9	—	5
34–36·9	—	2
37 and over	—	3
Total	100	100

287

Plot these data as cumulative frequency distributions and estimate graphically the medians and quartiles. Comment on the accuracy of your results. *(I.T.)*

40. At the end of 1975 there were 488 Trades Unions in the United Kingdom having the following membership.

Number of members	Number of Unions
Under 100	73
100 under 500	133
500 under 1 000	52
1 000 under 2 500	68
2 500 under 5 000	44
5 000 under 10 000	30
10 000 under 15 000	11
15 000 under 25 000	17
25 000 under 50 000	20
50 000 under 100 000	15
100 000 under 250 000	14
250 000 and over	11

SOURCE: Dept. of Employment

(a) Given that there are 4000 members (to the nearest thousand) in the 73 unions in the first group, find the appropriate lower limit to this group. Similarly, in the last group there are 7 264 000 members; calculate the upper limit for this group to the nearest thousand.

(b) Find the average size of union and the quartile deviation for this distribution.

41. The age distribution of the membership of a professional association is made up as follows.

Age	Number of Members
25–29	15
30–34	180
35–39	560
40–44	682
45–49	471
50–54	306
55–59	292
60–64	105
65–69	19

Find the mean age of the membership and the standard deviation of the distribution of the ages. Assuming that the ages are distributed evenly through each of the groups, what percentage of the membership is between the age of 32 years and 57 years?

42. The table shows the percentages of pupils in schools in England and Wales who took milk and dinners on a selected day in autumn in each of the years quoted. Calculate the average percentage of those who had milk and of those taking dinner. Calculate the standard deviation for each of the distributions and then compare their dispersion by using the coefficient of variation.

Year	1966	1967	1968	1969	1970	1971	1972
Percentage taking milk	68·4	69·5	70·1	70·0	67·9	59·8	64·0
Percentage taking dinner	80·1	79·2	91·4	91·4	91·2	94·6	94·7
	1973	1974	1975	1976			
	66·1	70·1	70·1	69·1			
	94·8	94·8	93·4	93·6			

SOURCE: Dept. of Education and Science

43. From the following table compare the consulting habits of men in the different age-groups, by deriving from the figures whatever additional statistics seem to you most suitable for the purpose. Comment on your choice of statistics and on your results.

Table showing the percentage distribution of male patients in 10 General Practices according to the number of consultations with their doctors.

| Age-group | Number of times consulting in the year | | | | | | | Total |
	0	1	2	3	4	5–9	10 or more	
0–	27·6	16·6	12·8	9·9	7·7	17·9	7·5	100·0
15–	41·4	15·6	10·8	8·3	6·4	12·8	4·7	100·0
45–	41·1	12·5	8·6	6·6	5·5	14·3	11·5	100·0
65 and over	33·5	11·0	7·0	7·8	5·4	17·9	17·4	100·0
All ages	36·8	14·8	10·6	8·3	6·5	15·0	8·0	100·0

SOURCE: General Register Office: General Practitioners' Records, Studies on Medical and Population Subjects No. 9, H.M.S.O., 1965

44. Using the data of Question 46 compare the arithmetic mean age of teachers with first class Honours degrees with that for all other teachers, and comment on your results. (*L.G.I.*)

45. Explain carefully what you understand by the 'mode' and the 'geometric mean'. In what circumstances might their use be appropriate? Why are they rarely used? Show, with examples, how you would estimate the mode of a grouped frequency distribution and how you would compute the geometric mean of a set of observations when weights are to be attached to each observation.

(L.G.I.)

46. The following table gives the age distribution of male teachers in maintained grammar schools in England and Wales who have degrees in Science and Mathematics. Compare the advantages of two different diagrams (other than cumulative frequency diagrams) that might be used to illustrate these data. Draw the diagram that in your opinion provides the better illustration.

Age Group	1st Class Honours	2nd Class Honours	Other Degrees	Total
20–24	23	119	77	219
25–29	68	380	533	981
30–34	34	206	356	596
35–39	57	165	203	425
40–44	104	335	250	689
45–49	160	314	336	810
50–54	147	282	369	798
55–59	72	155	260	487
60–64	41	72	153	266
65 and over	11	13	29	53

(L.G.I.)

47. From the table below, calculate the arithmetic mean number of sixth form pupils in boys' schools and girls' schools, respectively. State clearly any assumptions you make. Use a suitable graphic method to compare the distributions.

No. of sixth form pupils	No. of boys' schools	No. of girls' schools
under 20	28	151
20– 39	25	126
40– 49	7	26
50– 99	46	46
100–149	31	3
150–199	21	1
200–249	18	—
250+	4	—

(L.G.I.)

8. Explain the relationship between the frequency curve and the cumulative frequency diagram. What is a percentile? How can percentiles be used to measure dispersion? Draw the cumulative frequency diagram of the distribution of boys' schools in question 47. Use your diagram to determine the value of the median. Calculate also a measure of dispersion based on percentiles determined from your diagram   (*L.G.I.*)

9. Show how the arithmetic mean may be calculated from the grouped data by the 'short-cut' method, and explain how the method may still be used, if desired, when the class-interval is not constant throughout the tabulation. Use the method to compute the arithmetic mean of the following distribution:

Marks obtained	150–	170–	190–	200–	210–	230–	250–290
No. of candidates	16	67	194	268	182	96	23

(*L.G.I.*)

0. Use the data of Question 49 to estimate graphically:
   (*a*) the median of the distribution;
   (*b*) a measure of variability;
   (*c*) the mark that would be adopted as the 'critical level' if it were desired to eliminate the lower one-third of the candidates.

(*L.G.I.*)

1. The following figures represent the annual sales of a certain product over a 15-year period. Plot the original series and the trend line, using a 5-yearly moving average.

Year	Annual Sales (£)	Year	Annual Sales (£)
1965	5920	1973	7570
1966	5980	1974	7930
1967	6070	1975	7880
1968	6590	1976	7770
1969	7150	1977	8200
1970	7430	1978	7950
1971	7250	1979	8230
1972	7500		

2. Smooth the following series by means of a 5-year moving average. Represent the original series and the moving averages on the same graph.

Year	1868	1869	1870	1871	1872	1873	1874	1875
Imports (£M)	295	295	303	331	355	371	370	374

Year	1876	1877	1878	1879	1880	1881	1882
Imports (£M)	375	394	369	363	411	397	413

(I.S.A.

53.   Sales of Domestic Electricity (G.W.H. Ths)

Quarters	1	2	3	4
1975		19·9	14·2	24·5
1976	29·1	17·1	13·6	25·3
1977	28·6	18·8	14·9	23·6
1978	28·4	18·7		

Graph the above data and indicate the trend on your diagram

54. The figures below give the value in £m of exports of road vehicle
from Britain in the years 1977 and 1978. Plot the data given and
on the same graph put in the trend line obtained from a twelve
monthly moving average. Comment on this trend of the export o
road vehicles.

Year	Jan.	Feb.	Mar.	Apr.	May	June
1977	208	221	239	234	247	241
1978	211	279	270	288	252	265

Year	July	Aug.	Sep.	Oct.	Nov.	Dec.
1977	244	239	252	256	228	260
1978	253	260	254	288	224	—

55. The following table gives details of motor vehicle registrations ir
the U.K. between 1974 and 1977. Draw on one diagram the actua
figures together with the trend given by the 4-quarter centred
moving averages.

Motor Vehicle Registrations (Thousands)

Quarters	1974	1975	1976	1977
1	109·1	116·0	110·5	81·7
2	110·1	96·4	107·6	68·5
3	110·5	109·7	106·6	87·1
4	87·7	106·1	120·7	85·0

56. The following figures are the quarterly sales in millions of packet:
of a proprietary brand of food:

Quarters	1975	1976	1977	1978
		(million packets)		
1	48·2	46·2	62·0	69·1
2	52·6	63·0	57·2	51·9
3	87·2	89·8	83·8	89·0
4	73·0	87·0	76·0	78·0

Draw a graph to illustrate the given data and on the same diagram insert the trend. From the graph estimate the sales for 1954. (*R.S.A.I.*)

57. Using the method of moving averages calculate the average seasonal variation and the deseasonalised figures. Comment on the level of unemployment amongst females over the period 1969–72.

Unemployment in Great Britain (thousands)
Monthly average of unemployed females

Year	Quarter			
	1	2	3	4
1969	85	74	83	85
1970	85	79	90	93
1971	105	106	127	136
1972	144	133	147	132

SOURCE: Annual Abstract of Statistics

58. Use the method of moving averages to estimate the average seasonal variation in the following series:

Passenger movement by sea from the United Kingdom to the Irish Republic (thousands)

	1970	1971	1972	1973
1st Quarter	87	71	77	91
2nd Quarter	167	171	123	154
3rd Quarter	524	492	399	469
4th Quarter	80	82	92	99

SOURCE: Monthly Digest of Statistics

Use your estimates to make seasonal adjustments to the figures for the four quarters of 1973 and say what these seasonally adjusted figures indicate.

59. Calculate the average seasonal variation in the data below and prepare a new series with the seasonal element removed.

	1st Quarter	2nd Quarter	3rd Quarter	4th Quarter
1974	5244	6009	6078	6129
1975	6002	5660	6281	6480
1976	6763	7753	8208	8845
1977	9169	9601	9239	8987

60. Write explanatory notes on:
    (a) business charts
    (b) tabulation
    (c) averages (R.S.A.I.)

61.   U.K. Private Sector Bank Advances (£M)
                              1978

Jan.	Feb.	Mar.	Apr.	May	June	July	Aug.	Sept.
14772	14965	15032	15121	15360	15877	16594	16448	16215

Oct.	Nov.
16544	16662

Draw a graph of these advances and on the same diagram insert the trend.

62. Give an account of the main problems involved in constructing index numbers, considering in particular the construction of a retail price index. (I.T.)

63.   U.K. Index of Retail Prices 1975
                    (15th Jan. 1974 = 100)

Group	Weight	Index No.
Food	232	133·3
Alcoholic Drink	82	135·2
Tobacco	46	147·7
Housing	108	125·5
Fuel and Light	53	147·4
Durable Household Goods	70	131·2
Clothing and Footwear	89	125·7
Transport and Vehicles	149	143·9
Miscellaneous	71	138·6
Services	52	135·5
Meals bought out	48	132·4

SOURCE: Dept. of Employment

Compute the index number for (a) all items combined and (b) all items except 'transport and vehicles'.

64. Define the crude birth rate and the crude death rate. Explain the drawbacks of these rates, and describe briefly some of the more satisfactory measures of fertility and mortality. (L.G.F.)

65. From the following calculate the weight and index for Meat and Bacon.

Item	Weight	Index
Bread and Cakes	37	142
Meat and Bacon	—	—
Fish	8	125
Butter and Margarine	12	112
Milk, Cheese and Eggs	40	127
Tea, Coffee etc.	14	110
Sugar and Preserves	27	141
Vegetables	24	165
Fruit	15	125
Other Food	12	130
	250	Total Index = 136

66.

Index of Industrial Production
(1970 = 100)

Group	Group Indices (1970)	Weights
Glass	121	10
Bricks	100	17
Metal, ferrous	74	43
Metal, non-ferrous	92	14
Leather Goods	93	3
Paper and Printing	96	64
Food	101	52
Drink and Tobacco	121	32
Engineering	102	319
Vehicles	95	73

SOURCE: C.O.S.

Calculate an index number of production for the total of the ten sub-groups.

67. Discuss critically the difference between the present index of retail prices and the interim index. (R.S.A.A.)

68. Write a short report on the purpose and the use of index numbers, illustrating your answer by reference to a particular example with which you are familiar. (R.S.A.A.)

69. Calculate a weighted average of price relatives index for 1978 (base 1974 = 100) from the following.

	Price (£) Per Ton		
Raw Material	1974	1978	Weight
A	18	24	9
B	26	27	5
C	16	23	7
D	9	11	10
E	15	17	8
F	3	6	7

70.

Value of Imports–Indices
1970 = 100

Quarters	1	2	3	4
1971	102·8	106·0	106·5	105·6
1972	106·3	107·4	110·8	115·8
1973	122·2	131·8	144·7	159·5
1974	190·2	211·3	216·2	222·3

Explain the meaning of 1970 = 100.

Re-calculate these indices on the basis average 1971 = 100.

71. Calculate the median and the quartile deviation of the following distribution. In what circumstances is the quartile deviation likely to be more appropriate as a measure of dispersion than the standard deviation?

Age Distribution of U.K. Population
(age under 85 years) 1976

Age (years)	No. of People (Millions)
under 5	3·7
5 under 15	9·1
15 under 30	12·3
30 under 45	9·9
45 under 65	12·9
65 under 75	5·1
75–84	2·3

72. What are the advantages of the standard deviation as a measure of dispersion as compared with the range and the quartile deviation? Calculate the standard deviation of the following

distribution of times (to the nearest minute) taken by 100 conductors to 'pay in' at their garage at the end of a tour of duty.

Time taken to 'pay in' (minutes)	Number of Conductors
3	15
4	37
5	30
6	14
7	4
Total	100

(*I.T.*)

3. Calculate the mean and standard deviation of the following distribution.

Weight of article (grams)	Number of articles
0 under 8	98
8 under 16	82
16 under 24	56
24 under 32	44
32 under 40	40
40 under 48	22
48 under 56	13

4. From the data tabulated below calculate the arithmetic mean and standard deviation of daily earnings:

Gross daily earnings £	Number of men
8–	6
10–	15
12–	27
13–	31
14–	22
16–	16
18–20	3

5. Define:
(*a*) Range.
(*b*) First Quartile.
(*c*) Coefficient of Variation.

Calculate each of the above statistics for the sample of twelve values given below:

$$7, 5, 9, 0, 9, 4, 6, 6, 1, 2, 5, 3.$$

76. The following series of data gives the times taken in minutes b
    clerks to fill up a case record summary.
    Calculate:
    (a) the mean time to complete a summary;
    (b) the standard deviation of the time.
    From your results state what you think would be a reasonabl
    maximum time to allow for the completion of one summary.

No. of minutes	1	2	3	4	5	6	7	8	9	10	
No. of clerks	2	3	5	10	15	30	25	15	10	5	Total:12(

    (I.S.A

77. Define three measures of dispersion.
    Explain how the standard deviation is calculated for bot
    ungrouped and grouped data.
    What is meant by 'skewness' and how can it be measured?
    (I.S.I

78. Explain the meaning of the following: mode; coefficient of varia
    tion; interpolation.   (I.S.A

79. Describe three measures of dispersion and state carefully th
    circumstances in which each would be used.   (R.S.A.I

80. The data set out in the table below relate to twelve bus route
    Investigate, by means of scatter charts, the relationship (a) betwee
    stops per km and gear changes per km, and (b) between stops p
    km and scheduled speed. Comment on what your diagrams revea

Route	Average Number of Stops per km	Average Number of Gear Changes per km	Average Scheduled Speed (km/h)
1	3·52	10·93	15·5
2	5·15	15·80	11·3
3	3·98	14·96	11·5
4	1·98	6·43	16·9
5	1·85	7·18	14·7
6	2·73	6·68	16·7
7	2·86	8·19	15·2
8	2·69	7·54	15·6
9	4·91	12·97	11·2
10	5·24	15·48	12·2
11	3·93	11·07	12·7
12	3·82	8·60	14·0

    (I.T

81. The following data are the packed cell volume in millimetres (X) and red blood cell count (Y) of 10 dogs.

X	42	45	56	48	42	35	58	40	39	50
Y	6·30	6·53	9·52	7·50	6·99	5·90	9·49	6·20	6·55	8·72

   (a) Fit a linear regression of Y on X.
   (b) Estimate the red blood cell count for a packed cell volume reading of 52 millimetres.   (I.S.I)

82. The death rate amongst male children from birth to under five years of age for the period of 1964–76 in the United Kingdom is given below. Plot these data on a graph taking 1964 = 0 on the horizontal axis. Calculate the regression line of mortality rate on time and put this 'line of best fit' on your graph. Estimate the mortality rate in 1982 for this age group if the same trend continues.

Year	64	65	66	67	68	69	70	71	72	73
Death rate/1000	5·7	5·3	5·3	4·8	4·9	4·8	4·8	4·7	4·4	4·2

74	75	76
4·0	3·7	3·5

SOURCE: Annual Abstract of Statistics

83. (a) Calculate the product-moment coefficient of correlation for the following data.

Sales of Ice-cream £(100)	156	147	182	113	174	118	148	162
Temperature (°F)	60	59	72	51	71	52	58	65

   (b) What is the purpose of calculating correlation and what does its value indicate in respect of the above data?

84. The following are the gestation times and birth weights of 10 infants:

Gestation Time (days)	Birth Weight (kg)
240	2·95
250	2·50
255	3·18
260	4·09
270	4·09
275	3·63
280	3·18
285	4·54
290	4·31
310	4·77

299

Calculate the correlation coefficient between gestation time and birth weight, and comment on your result. (*L.G.*)

85. Define a measure of rank correlation.

Ten students obtained the following marks in Mathematics and English:

Student	1	2	3	4	5	6	7	8	9	10
Mathematics	60	89	42	94	92	29	86	97	80	74
English	62	69	49	70	68	74	40	82	87	28

Calculate the rank correlation coefficient between the two sets of marks. (*I.S.I.*)

86. Plot a scatter diagram for the following points and on the graph draw the line of best fit.

y	13·0	12·0	16·5	16·0	17·5	19·5	19·0	21·0
x	14·5	17·0	16·0	17·5	19·0	20·5	20·0	20·5

(*I.S.A.*)

87. Explain what is meant by correlation and give examples of data where (*a*) direct correlation, (*b*) indirect correlation might be expected.

Calculate the correlation coefficient for the following indices relating to production and price of a certain commodity:

Production *x*	92	96	103	108	109	108	96	103	109	103
Price *y*	110	111	94	93	89	84	100	106	87	94

(*R.S.A.I.*)

88. Calculate the product-moment coefficient of correlation for the following data.

Year	Coal Supply (10000 tons)	Number of Wage Earners at Colliery (Thousands)
1967	336	401
1968	322	350
1969	295	311
1970	279	290
1971	291	284
1972	240	271
1973	253	257
1974	215	245

89. Explain the meaning of three of the following:
(*a*) skewness, (*b*) the mode, (*c*) correlation, (*d*) the moving average. (*R.S.A.I.*)

90. The following table gives the number of criminal convictions in thousands and the number unemployed in millions, each year for a period of 10 years.

Number convicted (000s)	9·5	7·9	8·1	7·9	7·3	7·4	7·2	8·3	8·8	10·5
Number unemployed (mlns.)	2·3	1·3	1·2	1·4	1·2	1·3	1·3	2·5	2·7	2·8

Calculate the coefficient of correlation and explain the meaning of your results. (*R.S.A.I.*)

91. Show that the number of combinations of n objects, taken r at a time, is:

$$_nC_r = \frac{n!}{(n-r)!\,r!}$$

Show that $_nC_r + {_nC_{r+1}} = {_{n+1}C_{r+1}}$. (*I.S.I.*)

92. Three gamblers, **A**, **B**, **C**, take turns to throw an unbiased six-sided die, the winner being the first to throw a six. Show that the chances of **A**, **B**, **C** winning are as $36:30:25$, respectively.

If now rules are changed so that **A** wins if he throws a five or a six, **B** wins if he throws a four, five or six and **C** wins if neither **A** nor **B** has already won, show that **A**, **B**, and **C** now have an equal chance of winning. (*I.S.I.*)

Random Sampling Numbers

Instruction on how to use this table are given on page 117.

78 41	11 62	72 18	66 69	58 71	31 90	51 36	78 09	41 00
70 50	58 19	68 26	75 69	04 00	25 29	16 72	35 73	55 85
32 78	14 47	01 55	10 91	83 21	13 32	59 53	03 38	79 32
71 60	20 53	86 78	50 57	42 30	73 48	68 09	16 35	21 87
36 30	15 57	99 96	33 25	56 43	65 67	51 45	37 99	54 89
09 08	05 41	66 54	01 49	97 34	38 85	85 23	34 62	60 58
02 59	34 51	98 71	31 54	27 85	23 84	49 07	33 71	17 88
20 13	44 15	22 95	98 97	60 02	85 07	17 57	20 51	01 67
36 26	70 11	63 81	27 31	79 71	08 11	87 74	85 53	86 78
00 30	62 19	81 68	86 10	65 61	62 22	17 22	96 83	56 37
38 41	14 59	53 03	52 86	21 88	66 87	85 59	14 90	74 87
18 89	40 84	71 04	09 82	54 44	94 23	83 89	04 59	38 29
34 38	85 56	80 74	22 31	26 39	65 63	12 38	45 75	30 35
55 90	21 71	17 88	20 08	57 64	17 93	22 34	00 55	09 78
81 43	53 96	96 88	36 86	04 33	31 40	18 71	96 00	51 45
59 69	13 03	38 31	77 08	71 20	23 28	92 43	92 63	21 74
60 24	47 44	73 93	64 37	64 97	19 82	27 59	24 20	00 04
17 04	93 46	05 70	20 95	42 25	33 95	78 80	07 57	86 58
09 54	99 70	03 41	39 91	27 40	72 05	55 17	08 80	15 19
19 55	42 30	27 05	27 93	78 10	69 11	29 56	29 79	28 66
46 69	28 64	81 02	41 89	12 03	31 20	25 16	79 93	28 22
28 94	00 91	16 15	35 12	68 93	23 71	11 55	64 56	76 95
59 10	06 29	83 84	03 68	97 65	59 21	58 54	61 59	30 54
41 04	70 71	05 56	76 66	57 86	29 30	11 31	56 76	24 13
09 81	81 80	73 10	10 23	26 29	61 15	50 00	76 37	60 16

```
91 55   76 68   06 82   05 33   06 75   92 35   82 21   78 15   19 43
82 69   36 73   58 69   10 92   31 14   21 08   13 78   56 53   97 77
03 59   65 34   32 06   63 43   38 04   65 30   32 82   57 05   33 95
03 96   30 87   81 54   69 39   95 69   95 69   89 33   78 90   30 07
39 91   27 38   20 90   41 10   10 80   59 68   93 10   85 25   59 25

89 93   92 10   59 40   26 14   27 47   39 51   46 70   86 85   76 02
99 16   73 21   39 05   03 36   87 58   18 52   61 61   02 92   07 24
93 13   20 70   42 59   77 69   35 59   71 80   61 95   82 96   48 84
47 32   87 68   97 86   28 51   61 21   33 02   79 65   59 49   89 93
09 75   58 00   72 49   36 58   19 45   30 61   87 74   43 01   93 91

63 24   15 65   02 05   32 92   45 61   35 43   67 64   94 45   95 66
33 58   69 42   25 71   74 31   88 80   04 50   22 60   72 01   27 88
23 25   22 78   24 88   68 48   83 60   53 59   73 73   82 43   82 66
07 17   77 20   79 37   50 08   29 79   55 13   51 90   36 77   68 69
16 07   31 84   57 22   29 54   35 14   22 22   22 60   72 15   40 90

67 90   79 28   62 83   44 96   87 70   40 64   27 22   60 19   52 54
79 52   74 68   69 74   31 75   80 59   29 28   21 69   15 97   35 88
69 44   31 09   16 38   92 82   12 25   10 57   81 32   76 71   31 61
09 47   57 04   54 00   78 75   91 99   26 20   36 19   53 29   11 55
74 78   09 25   95 80   25 72   88 85   76 02   29 89   70 78   93 84
```

Answers to the Exercises

Exercise 2.1, page 11

1. 4·76   2. 31   3. 0·85   4. 5760   5. 6000
6. 800   7. 10000   8. 0·004   9. 0·0040   10. 201
11. 200   12. 1000

Exercise 2.2, page 12

1. Correct 2. One   3. Correct 4. One   5. One
6. One   7. Correct

Exercise 2.3, page 12

1. $2\cdot73 \pm 0\cdot005$ metres   2. $12 \pm 0\cdot5$ metres   3. $10\cdot2 \pm 0\cdot05$ sec
4. Correct   5. Correct

Exercise 2.4, page 14

1. $60\cdot7 \pm 0\cdot1$ 2. $6\cdot91 \pm 0\cdot01$ 3. $164\cdot6 \pm 0\cdot1$
4. $2\cdot02 \pm 0\cdot01$ 5. $67\cdot11 \pm 0\cdot555$ 6. $32\cdot11 \pm 0\cdot01$
7. $18\cdot7 \pm 0\cdot55$ 8. $10\cdot0 \pm 0\cdot2$ 9. $40\cdot046 \pm 0\cdot069$
10. $0\cdot01982 \pm 0\cdot00026$ 11. $3\cdot97 \pm 0\cdot01$ 12. $10\cdot02 \pm 0\cdot11$
13. $76\cdot20 \pm 0\cdot66$ 14. $£1\cdot81 \pm £0\cdot02$ 15. $20\cdot3 \pm 0\cdot2$
16. $56\cdot8 \pm 1\cdot5$ 17. $9\cdot8 \pm 0\cdot1$ 18. 305 ± 8
19. $23\cdot45 \pm 0\cdot06$ 20. 64656 ± 501 21. 454 to 470
22. 39 to 42   23. $3\cdot97 \pm 0\cdot01$ 24. (i) 99; (ii) 102
25. 52778812 ± 2351673 26. 1555 ± 55
27. 143102 ± 1205; $128576\cdot5 \pm 1088\cdot5$ 28. $2\cdot12 \pm 0\cdot17$

Exercise 2.6, page 23

1. (i) 0·0065; 0·65   (ii) 0·017; 1·7   (iii) 0·029; 2·9
   (iv) 0·019; 1·9   (v) 0·031; 3·1   (vi) 0·024; 2·4
   (vii) 0·014; 1·4   (viii) 0·048; 4·8   (ix) 0·018; 1·8

2. (i) $\dfrac{x_1}{x_2}\left(\dfrac{2e_1}{x_1}+\dfrac{e_2}{x_2}\right)$; $(2r_1+r_2)$; $(2p_1+p_2)$

   (ii) $\dfrac{4x_1}{x_2}\left(\dfrac{e_1}{x_1}+\dfrac{2e_2}{x_2}\right)$; (r_1+2r_2); (p_1+2p_2)

   (iii) $\dfrac{2x_1}{x_2}\left(\dfrac{e_1}{x_1}+\dfrac{e_2}{x_2}\right)$; (r_1+r_2); (p_1+p_2)

3. $\dfrac{x_1}{x_3}\left(\dfrac{e_1}{x_1}+\dfrac{e_3}{x_3}\right)+\dfrac{x_2}{x_3}\left(\dfrac{e_2}{x_2}+\dfrac{e_3}{x_3}\right)$

4(a) £(135000 ± 1165)   (b) 0·0086

Exercise 2.7, page 25

(i) $(1+2a-3b)$ (ii) $\left(1+\dfrac{a}{2}-\dfrac{b}{2}\right)$ (iii) $\left(1+\dfrac{a}{2}+3b\right)$

(iv) $\left(2+\dfrac{a}{2}+\dfrac{b}{2}-c\right)$ (v) $\left(1+3a+\dfrac{b}{3}-\dfrac{c}{4}\right)$

Exercise 3.1, page 31

5(a) Discrete   (b) Discrete   (c) Continuous   (d) Discrete
 (e) Discrete   (f) Continuous   (g) Discrete   (h) Continuous
13(a)(i) Continuous   (ii) Discrete   (iii) Continuous   (iv) Discrete
 (v) Discrete

Exercise 4.1, page 44

4(a) 90, 90, 180   (b) 30, 60, 90, 180   (c) 30, 60, 60, 90, 120
 (d) 5, 40, 75, 100, 140   (e) 69·8, 134·6, 155·6

Exercise 4.2, page 45

8. 6440, 8804, 2456.   9. 136·8°, 36°, 10·8°, 14·4°, 18°, 43·2°,
   28·8°, 14·4°, 43·2°   12. (i) 55°, 49°, 177°, 55°, 24° (iii) 5·57 cm
13. 20·6°, 13·3°, 2·1°, 44°, 220·4°, 38·5°, 21·1°
14(a)(i) 23·1   (ii) −37·8%   (iii) 29·6   (iv) 96·1   (v) 70·0
   (b) 115·8°, 121·7°, 110·9°, 11·6°
18. 4·48 cm; 4·19, 0·35, 0·21, 0·09

Exercise 5.1, page 58

13(i) 295°, 36°, 19°, 10°   (iii) 5·26 cm

Exercise 6.1, page 78

4(a) 1, 2, 3, 4, 5, 6, 7; 0·5–1·5, 1·5–2·5,...6·5–7·5
 (b) $\frac{1}{2}$, $1\frac{1}{2}$, $2\frac{1}{2}$, $3\frac{1}{2}$, $4\frac{1}{2}$, 6, $8\frac{1}{2}$; 0–1, 1–2,...5–7, 7–10
 (c) 1, 2, 3, 4, 5, 6, 7, 8, –; 0·5–1·5...8·5–
 (d) 3, $7\frac{1}{2}$, $12\frac{1}{2}$, 20, 30, 45; 1–5, 5–10, 10–15, 15–25, 25–35, 35–55
 (e) 5, 15, $22\frac{1}{2}$, $27\frac{1}{2}$, –; 0–10, 10–20,...30–
 (f) $2\frac{1}{2}$, $7\frac{1}{2}$, $12\frac{1}{2}$, $17\frac{1}{2}$, $22\frac{1}{2}$, $27\frac{1}{2}$, $32\frac{1}{2}$, 0 and less than 5, 5 and less than
   10,...30 and over
 (g) 3, 8, 13, 18, 23; 0·5–5·5, 5·5–10·5, 10·5–15·5, 15·5–20·5, 20·5–25·5
20. 3, 8, 13, 20·5, 30·5, 45·5
23. 165, 195, 225, 255, 285, 330, 450.

Exercise 7.1, page 94
5. 9·42; 10·22; 9·52; 9·0; 7·34

Exercise 7.2, page 102
1. 20, 5, 3·5, 3, 2·75, 2·5, 1, 0·5, 0·1
2. 1 cm, 44·1 cm, 2·9 km, $6·7 \times 10^4$ km, $1·10 \times 10^4$ km, $1·1 \times 10^5$ km, $1·24 \times 10^6$ km, $8·97 \times 10^7$ km, $2·14 \times 10^8$ km, $3·31 \times 10^8$ km, $2·97 \times 10^8$ km, $3·31 \times 10^9$ km, $6·69 \times 10^9$ km, $5·52 \times 10^{10}$ km

Exercise 10.1, page 119
2. No   3. Yes   5. Usually yes   6. Usually yes   14. 71431 ± 496

Exercise 11.1, page 125
1. 748·8
2(a) 13·1   (b) 418·7   (c) 3·75   (d) $31\frac{1}{4}$
 (e) 90   (f) $4\frac{19}{40}$ (g) 406·3
3(a) $3^2 \times 7$; 714   (b) $2^2 \times 3$; 254·4   (c) 3×7; 205·8
 (d) 111; 1198·8

Exercise 11.2, page 126
1. 41   2. 64   3. 73   4. 63   5. 56   6. 25

Exercise 11.3, page 127
1. 6   2. 9   3. 41   4. 30   5. £11   6. 6

Exercise 11.4, page 128
1. 13·95   2. 24·80   3. 51·68   4. 69·68

Exercise 11.5, page 128
1. 981·2   2. 25 yr 3 mth   3. 50·25, 48·5, 52·5
4. 321·5; 1952(4th) 16·7; 1953(2nd) 2   5. 393; 387; 399
6. 57; 50; $54\frac{1}{9}$ 7. $7\frac{11}{14}$ 8. 15, 14·5, 14   9. £8·61
10. £77·5, £80, £70, mean £77·98   11. 11·25   12. 61·1, 62, 62
13. 4, 3, 5, 4·34   14. $\sqrt{ab}$

Exercise 12.1, page 132
1. 5·87   2. 6·2   3. 3·3   4. 30·44   5. 162·35

Exercise 12.2, page 133
1. 18·1   2. 39·65   3. 36·85   4. 49·03

Exercise 12.3, page 135
1. 13·2   2. 26·25   3. 40·0 and 39·7   4. 12·2 to 17·2   5. 13·8

Exercise 12.4, page 142
1. 13·3   2. 26·5   3. 39·8 and 39·3   4. 14·8   5. 14·2

Exercise 12.5, page 146
1. 13·5   2. 27·2   3. 42·6 and 42·3   4. 14·4   5. 15·2

Exercise 12.6, page 148
1. 9·4, 9·7, 10·15   2. 3·77, 3·4, 2·8   3. 13·9, 11·8, 3·5

Exercise 12.7, page 148
1. 21·6   2. 44·8, 49·8, 48·2   3. £56·5   4. 327, 313, 57
5. 15, 55, 71, 32   6. 148   7. 155, 25   8. £24, £41·9
   (a) B and D, £12·10 and £2·10   (b) A and C, £9·9 and £4·40
9. 42·5   10. 43 mm, 11·0 mm   11. 5·50, 5·27
12. 62·1, (a) 67%   (b) 92
13. 49·5; 46·4; grouped nearer to 50 than 41, Paper II harder.
14. £522, £12·73; £257; £13·53   15. (i) 164·2   (ii) 164·9
16. 5·3%   17. £1955, £1349, £2756
18. (i) 20·02; (ii) 19·94 to 20·098 mm; 20·025 mm; 5 components

Exercise 13.1, page 163
1. (a) 5·4, 6·2, 6·6, 6·8, 8·0, 8·8, 9·6
   (b) 18·8, 16·6, 13·8, 12·0, 10·2
   (c) 21·66, 22·88, 24·00, 25·00, 25·66, 26·28, 26·72, 27·14, 27·58,
       27·94, 28·30, 28·76
2. 102, 105·7, 111, 109, 106·3, 104·7, 106, 106·3, 104·3, 102·3
3. 20, 20, 20, 20, 20, 20, 20, 21, 22, 23, 24, 25, 26, 27, 28, 29
4. 195·9, 196·0, 198·9, 201·8, 203·8, 202·5, 196·6, 190·8
5. 256, 252, 243, 256, 260, 270, 276, 281, 284
6. 88, 103, 118, 110, 96, 81, 70, 60, 48, 40
7. 10060, 10398, 10675, 11039, 11306, 11603, 12021, 12533, 13114
12. −55, −50, −35
13. 114, 118, 122, 125, 128, 131, 134, 136, 137, 138, 139, 143
14. 6 yrs; 1·6   16. 17·8–18.
17. 105·6; 112·9; 144·4; 160·5

Exercise 14.1, page 170

1. **A** 150, 67   **B** 138, 72   **C** 73, 136   **D** 105, 95
   E 86, 117   **F** 68, 147   **G** 374, 27   **H** 94, 106
   I 99, 101   **J** 131, 77
2. **A** 89   **B** 95   **C** 94   **D** 81   **E** 58

Exercise 14.2, page 172

1.   **A** 96, 101, 102, 100, 101   **B** 87, 121, 91, 128, 73
    C 101, 103, 79, 105, 112
2(a) **A** 100, 120, 167, 237, 351   **B** 100, 102, 107, 114, 125
    C 100, 103, 103, 100, 99
 (b) **A** 29, 34, 48, 68, 100   **B** 80, 82, 86, 91, 100
    C 101, 104, 104, 101, 100

Exercise 14.3, page 176

1(a) 125   (b) 134   (c) 65·7   2. 116   3. 108

Exercise 14.4, page 180

1(a) (i) 100, 85·4, 87·9, 73·9, 66·2
    (ii) 151·0, 128·8, 132·7, 111·5, 0
    (iii) 120·9, 103·2, 106·3, 89·4, 80·1
    (iv) 100, 85·4, 102·9, 84·1, 89·7
 (b) (i) 100, 102·9, 105·8, 108·6, 111·5, 114·2, 116·2, 121·1, 121·8, 122·6
    (ii) 81·6, 84·0, 86·3, 88·6, 90·9, 93·1, 94·8, 98·8, 99·4, 100
    (iii) 88·9, 91·5, 94·1, 96·6, 99·1, 101·5, 103·3, 107·7, 108·3, 109·0
    (iv) 100, 102·9, 102·8, 102·7, 102·6, 102·4, 101·8, 104·2, 100·6, 100·6
2(a) 100, 103, 101, 102, 106, 108, 109, 112, 116
 (b) 85·5, 87·9, 86·3, 87·1, 90·6, 92·4, 93·3, 96·2, 100
3. 224, 221   4(a) 144   (b) 163   (c) 176   (d) 135   (e) 161
5. **A** 63·1   **B** 47·2   **C** 57·1   **D** 35·1   **E** 59·0   **F** 50·6
   G 65·0
6. Male: 6·7, 0·46, 0·41, 0·76, 1·0, 1·2, 2·5, 7·6, 22·4, 54·6, 128, 258
   Female: 5·1, 0·33, 0·28, 0·36, 0·54, 0·90, 1·9, 4·6, 11·5, 32·5, 91·3, 226
8. 144   9. 114
10(a) 10   (b) 13, 0·8, 3·3, 3·3, 3·75, 10·7, 24, 50   (c) 9·3   11. 114
12. 95·21   13. **B** greater than **A** by 1·653 deaths per thousand
14. 69·4   15. 101·0, 99·5
16. 89·3   17. 9·72   18(a) 160·4   (b) 186
19. 19·16, 6·556, 7·040   20. 13·6, 24·0
21(a) £12.19, £12.55, £12.30, £12.55   (b) 90·91, 122·73, 81·48, 77·07
22. 12·34, 12·85   23. 100, 104, 108·2, 115·7   24. 14

Exercise 15.1, page 189
1. 2·6   2. 2·8   3. 5·5   4. 3·6

Exercise 15.2, page 190
(a)(i) 2·65   (ii) 5·66, 4·00, 2·83, 3·32
(b)(i) 2·74   (ii) 3·39, 4·85, 6·60
(c)(i)19·14   (ii) 20·4, 19·8, 19·4, 20·8
(d)(i) 3·85   (ii) 3·98, 7·99, 11·7, 6·31
(e)(i) 1·60   (ii) 10·1, 4·31, 2·56

Exercise 15.3, page 192
1. 4·00, 2·24, 3·16, 1·86, 3·64, 3·09   2. 4·06

Exercise 15.4, page 195
1. (Exercise 12.2): 1. 6·53, 7·67   2. 9·56, 12·4   3. 16·1, 18·7
   4. 15·3, 18·4
   (Exercise 12.3): 1. 4·66, 5·92   2. 10·9, 13·1
   3. 7·31, 7·63 and 11·14, 11·22   4. 5·04, 5·97   5. 4·41, 5·41
2(a)(i). 46·4   (ii). 28·4   (b)(i). 46·2   (ii). 27·9
  (c)(i). 46·4   (ii). 27·9   (d)(i). 46·9   (ii). 27·3
3. Mean: 20·0 18·0 Mean Deviation: 9·7 10·8 σ: 11·3 12·1
4. 153 to 183 cm   (a) 680   (b) 950   5. 1st group
6(a) 16%   (b) 84%   (c) $2\frac{1}{2}$%

Exercise 15.5, page 199
1. 24, 4·7, −0·05   2. 49, 11·1, −0·07   3(a) 55, 14, −0·23
3(b) 55, 12·4, −0·23 4. 24, 4·54, +0·05
5. 23, 2·85, −0·26

Exercise 15.6, page 199
1(a) 2   (b) 1·004   3. 1070 hr; 306 hr   4. 5·4, 0·37 sec
5. 0·55, 0·86   6. 6·6, 3·7   7. £12.31, £3.89
8. 4·3, 0·387   9. −0·0935
10. 7·5, 0·616   11. 1·636   12. 29·09, 11·37
13. 5·19, 0·23   14(a) 29·5, 3·39   (b) 22·1, 2·26
15(a) 210·67   (b) 212·7   (c) 10·44   (d) 14·24
16(a) 100·5   (b) 10·1   (c) 13·5   17. (i) 48·2, 15·3   (ii) 241, 79
18. (i) 8438   (ii) 2285, −117, −361, 448, −1364, −1181, −1235, −1404,
       502, −121, 1034, 1515   (iii) 964
19. (i) 16   (ii) 16   (iii) 16   (iv) 4·94   (v) 26
20. (i) £3.95 thousand. £1.71 thousand. £3.84 thousand.
     £1.66 thousand   (ii) Grocers with meat

21. 30·08, 10·28　(i) 19·80, 40·36　(ii) 9·52, 50·64
22. (i) 699·5　(ii) 899·5　(iii) 100　(iv) 703　(v) 149·3　(vi) 100
(vii) 7·5%

Exercise 15.7, page 207
1. 50, 20　　　　　2. (i) 0　　(ii) 1; D,A,C,B
3. 56·25, 25; **A** 34·4, **B** 53·6, **C** 87·2
4.

	A	**B**	**C**	**D**	**E**	
Maths.	64	84	66	52	65	
Eng.	64	52	56	80	64	
Av. Mark	72	$74\frac{1}{2}$	68	76	72·5	**D**
No, **B** does	Pcl. Weights 8/7					

Exercise 15.8, page 210
1. $\dfrac{nM-S}{4}$　　6. 44, 4·04　　7. $10d^2$, $14d^2$　　8. 2·66

9(a) $\dfrac{n+1}{2}\sqrt{\dfrac{n^2-1}{12}}$,

(b) $\dfrac{(n+1)(2n+1)}{6}$, $\sqrt{\dfrac{(n+1)(16n^3+14n^2-19n-11)}{180}}$

10. $\dfrac{2^n-1}{n}$　　11. $\dfrac{1}{3n^2}\left[2^{2n}(n-3)+6\times 2^n-(n+3)\right]$

12. $\dfrac{a^2}{n^2}\left[\dfrac{n(r^{2n}-1)}{(r^2-1)}-\left(\dfrac{r^n-1}{r-1}\right)^2\right]$

13. 5, 11　　　　14. 10, 3·742
15. $x_1^2+x_2^2+\ldots+x_n^2 = X_1^2+X_2^2\ldots+X_n^2-n\overline{X^2}$
16. 52·3, 50·8

Exercise 16.1, page 217
1. 5, 75, 3　　　2. 0·60, 39, -9　　3. $-2·0$, 35, 18
4. 0·82, 5, 37　　5. 1·7, 36, 6·5

Exercise 16.2, page 221
6. 61, 33

Exercise 16.3, page 225
9. 61　　10. 101, 91　　12. 2·6; 1·7　　14. **G**; **M**
17. **O** is farthest above the line

Exercise 16.4, page 231
1. $P = 1·6+0·69M$; $M = 0·59+0·88P$　　2. $X = 0·6Y+4·1$
3. $C = 81·52L-262·71$; $C = 161·2$　　4. $X = 24+0·4Y$

Exercise 16.5, page 234

1. $a \simeq 3{\cdot}5, b \simeq 2$ 2. $a = 64{\cdot}5, b = -1{\cdot}1$ 3. $a = 31, b = -1{\cdot}9$

4. $P = 19{\cdot}1 + 0{\cdot}18M$ 5. $y = \frac{1}{2} + \frac{2}{x}$ 6. $y = 3 + 2x^2, \ 8{\cdot}1, \ 3{\cdot}3$

7. $y = \frac{x^2}{5} + \frac{1}{2}$ 8. $\frac{x}{y} = \frac{1}{2}x + 1\frac{1}{2}$ 9. $y = \frac{2}{5}x^2 + 3\frac{1}{2}$

Exercise 17.1, page 238

1. $0{\cdot}99$ 2. $0{\cdot}77$ 3. $0{\cdot}96$ 4. $0{\cdot}604$, distinct correlation
5. $-0{\cdot}858$, high production goes with low prices.

Exercise 17.2, page 241

1. (1) 1 (2) $0{\cdot}88$ (3) $0{\cdot}95$ (4) $0{\cdot}68$ (5) $-0{\cdot}84$
2. $0{\cdot}58$ 3. $-0{\cdot}19, 0{\cdot}12, -0{\cdot}21$ 4. $-0{\cdot}91$
5(a) 1 (b) -1 (c) $-0{\cdot}03$

Exercise 17.3, page 242

1. $-0{\cdot}39$ 2. $0{\cdot}02$ 3. $-0{\cdot}833$ 4. $0{\cdot}897$ 5. $0{\cdot}45$
6. $+0{\cdot}77$, price of product tends to hold its relative position in stores
7. $0{\cdot}21$, low value, so price not a good guide.
8(a) (i) -1 and 1 (ii) 0 (b) $0{\cdot}97$ 9. $0{\cdot}8, 0{\cdot}8$

Exercise 18.1, page 247

1(a) $3/4$ (b) $2/3$ (c) $3/8$ (d) $1/2$ (e) $1/8$ (f) $5/8$ (g) $1/4$
2(a) $1/5$ (b) $1/(13)$ (c) $3/(13)$ (d) $1/3$ (e) $5/8$ (f) $1/2$ (g) $1/2$
   (h) $1/(40)$ (i) $6/(11)$
3(a) $1/9$ (b) $2/9$ 4. $(11)/(15)$ 5. 360
6(a) $1/5$ (b) $1/6$ (c) $1/3$ (d) $3/10$
7(a) $1/(36)$ (b) $1/12$ (c) $1/9$ (d) $1/(12)$ (e) $1/(36)$ (f) 0
8(a) $1/3$ (b) $3/7$ (c) $1/4$ (d) $3/7$ 9(a) $\dfrac{b}{b+r}$ (b) $\dfrac{r}{b+r}$

10. $7/(17), 5/8$ 11. $1/9$

Exercise 18.2, page 249

1. $4/(13)$ 2. $1/3$ 3. $(10)/(23)$ 4. $4/(13)$
5. $(17)/(24)$ 6. $3/5$ 7. $3/5$

Exercise 18.3, page 250

1. $1/(169)$ 2. $(30)/(169)$ 3. $1/(12)$ 4. $1/3^{12} = 1/531\,441$
5(a) $p_1 p_2$ (i) $1/6$ (ii) $1/(32)$ (iii) $1/(24)$ (iv) $9/(32)$ (v) $35/64$
   (b) $p_1(1-p_2)$ (i) $1/3$ (ii) $7/(32)$ (iii) $7/(24)$ (iv) $3/(32)$
       (v) $5/(64)$

(c) $(1-p_1)(1-p_2)$ (i) $1/3$ (ii) $(21)/(32)$ (iii) $7/(12)$ (iv) $5/(32)$
(v) $3/(64)$

(d) $(1-p_1)p_2$ (i) $1/6$ (ii) $3/(32)$ (iii) $1/(12)$ (iv) $(15)/(32)$
(v) $(21)/(64)$

6(a) $p_1(1-p_2)(1-p_3)(1-p_4)$ (b) $p_1p_2(1-p_3)(1-p_4)$
(c) $p_1p_2p_3(1-p_4)$ (d) $p_1p_2p_3p_4$

7. $1/(6561)$ 8. $5/(33)$ 9. $1/(7893600)$

Exercise 18.4, page 252
1. $0\cdot7$ 2. $0\cdot274$ 3. $0\cdot663$ 4. $0\cdot034$
5. $0\cdot464$ 6. $(19)/(27)$ 7. $(65)/(81)$ 8. $1/2$
9(a) $1/(12)$ (b) $5/(648)$ (c) $(31)/(216)$ 10. $2/9, 5/9, 7/9$
11. $1/(34720)$; $(3551)/(6944)$ 12. $(67)/(150)$; $(67)/(300)$

Exercise 19.1, page 256
1. $60; 42; 40320$ 2. $6, 9$ 3. $\dfrac{(13)!}{2!}$
4. 72 5. 986409 6. 3628800
7. 720 8. 20 9. 1680
10. 117600 11. 3991680 12. 326

Exercise 19.2, page 258
1. $60; 45360; 151200; 3360; 2520; 50400; 907200; 20160$
2. $\dfrac{16!}{5!6!3!2!}$ 3. $840; 120; 820$ 4. 110880 5. $\dfrac{21!}{(3!)^7}$
6. 3 7. 36 8. 1680 9. 60 10. 978

Exercise 19.3, page 259
1. 125 2. 3^{12} 3. 9×10^3 4. 64
5. 10000 6. 1176 7. 120

Exercise 19.4, page 260
1. 1023 2. 79 3. 287 4. 11 5. 31

Exercise 19.5, page 262
1. 4571591 3. $\dfrac{30!}{22!8!}$ 4. $\dfrac{60!}{60!20!}$ 5. $_{60}C_{40} \cdot _{20}C_{15}$
6. $_{17}C_2 \cdot _{14}C_3$ 7. $_{10}C_2$; $_{10}C_3$ 8. 24

Exercise 19.6, page 262
1. $\dfrac{1}{_{52}C_4}$ 2. $\dfrac{_{98}C_8}{_{100}C_{10}} = \dfrac{1}{110}$ 3. $\dfrac{_{14}C_8}{_{16}C_{10}} = \dfrac{3}{8}$ 4. $\dfrac{_9C_5}{_{10}C_6} = \dfrac{3}{5}$

5. $2/(11)$ 6. $1/(10)$ 7. $6/(85)$ 8. $(55)/(91)$

9. $7/(22)$ 10. $(35)/(726)$; $7/(99)$

11(a) ${}_{30}C_{10}$ (b) ${}_{20}C_6 \cdot {}_{10}C_4$ (c) ${}_{27}C_7$ 12. $2_5 C_3 \cdot 9 C_7$

13. 66; $(10)/(11)$ 14. ${}_{12}C_3 \cdot {}_{14}C_4$ 15. ${}_{10}C_3 \cdot 7 C_3$ 16. 120, 12

17. 100 18. 456 19. 60

20. $10 \times 11^2 \times 12^2 \times 13 \times 14$ 21. 168 22. $12 \times 8!$

23. $570\,240$ 24(a) 9 (b) 10 25. 2400; 1200 26. 192

27. $648\,000$ 28(a) (i) 8 (ii) 6 (b) $16\,170$; $12\,600$

Exercise 20.1, page 266

1. $1 + 6x + 15x^2 + 20x^3 + 15x^4 + 6x^5 + x^6$
2. $1 + 6x + 12x^2 + 8x^3$
3. $16x^4 - 96x^3 y + 216x^2 y^2 - 216xy^3 + 81y^4$
4. $p^8 + 8p^7 q + 28p^6 q^2 + 56p^5 q^3 + 70p^4 q^4 + 56p^3 q^5 + 28p^2 q^6 + 8pq^7 + q^8$
5. $32x^{10} - 240x^8 + 720x^6 - 1080x^4 + 810x^2 - 243$

Exercise 20.2, page 267

1(a) $x^4 + 4x^3 y + 6x^2 y^2 + 4xy^3 + y^4$
 (b) $1 - 5x + 10x^2 - 10x^3 + 5x^4 - x^5$
 (c) $x^4 - 4x^3 y + 6x^2 y^2 - 4xy^3 + y^4$
 (d) $x^5 + 10x^4 y + 40x^3 y^2 + 80x^2 y^3 + 80xy^4 + 32y^5$
 (e) $x^7 + 14x^6 + 84x^5 + 280x^4 + 560x^3 + 672x^2 + 448x + 128$
 (f) $6561x^8 - 34992x^7 y + 81648x^6 y^2 - 108864x^5 y^3 + 90720x^4 y^4$
       $- 48384x^3 y^5 + 16128x^2 y^6 - 3072xy^7 + 256y^8$

2(a) $84p^6 q^3$ (b) $66x^2$ (c) $330x^7 y^4$ (d) $252m^5 n^5$ (e) $1716a^7 b^6$

3(a) $1 \cdot 1041$ (b) $1 \cdot 3048$ (c) $2 \cdot 9860$ (d) $247 \cdot 0771$ (e) $15 \cdot 3695$

5(a) $\dfrac{220}{177\,147}$ (b) $\dfrac{5280}{177\,147}$ (c) $\dfrac{42\,240}{177\,147}$ (d) $\dfrac{28\,160}{177\,147}$ (e) $\dfrac{1320}{177\,147}$

6(a) $160x^3 y^3$ (b) 5th

7(a) $16x^4 - 24x^3 + \dfrac{27}{2}x^2 - \dfrac{27}{8}x + \dfrac{81}{256}$ (b) 3rd

8(a) 112 (b) -25 9(a) $\dfrac{105}{8}x^6$, 840 (b) $-\dfrac{160}{27}$

Exercise 20.3, page 271

1. $\dfrac{25}{72}$ 2. $\dfrac{35}{128}$ 3. $\dfrac{3942}{4096}$

4. $\dfrac{1}{28\,561}$, $\dfrac{48}{28\,561}$, $\dfrac{864}{28\,561}$, $\dfrac{6912}{28\,561}$, $\dfrac{20\,736}{28\,561}$

5. $\dfrac{79}{4096}$ 6. $\dfrac{16}{81}$; $\dfrac{32}{81}$; $\dfrac{8}{27}$; $\dfrac{8}{81}$; $\dfrac{1}{81}$ 7. $0 \cdot 83$ 8. $0 \cdot 058$

9. $0 \cdot 32\%$

Exercise 20.4, page 274

1(a) 1; $\sqrt{\frac{1}{2}}$ (b) 3/2; $\sqrt{3}/2$ (c) 5/3; $\sqrt{10}/3$

2. 48; $\frac{1}{4}$; $\frac{3}{4}$ 3. 16

4(a) $(2/5)^{11}$ (b) $(2/5)^{15}$ (c) 5 to 8   (d) 7 to 11

7. 0·00054   9. 27%   10. 7 to 13

11. 0·39   12. 34 to 46   13. 0·0038

14. 380 to 388; 0·81   15. $7\frac{1}{2}$–$12\frac{1}{2}$ 16. 0·26; 0·043

17. 100; 1/5; 4/5   18. 200±14   19. 0·84

20. $\frac{1}{4}$ or $\frac{3}{4}$ 21. $p_1 = 1/9$; $p_2 = 1/15$

22(a) 0·133   (b) 0·319   24. (17)/(625); (11)/(144)   25. 52

26. Heads: 0   1   2   3   4   5
    Expected frequency: 90 450 900 900 450 90, 2340

27. Heads: 0   1   2   3
    Expected frequency: 2061 6183 6183 2061

28. (i) 1/(16) (ii) 3/8 (iii) (11)/(16)   29. (i) 5/(16) (ii) (11)/(32)

30. 1/4, 1/5

Miscellaneous Exercises, page 278

1. (a) 7·5   (b) 45·5, 44·1   2. (a) 135 500 ± 1165   (b) 0·0086

14. 84·3°, 102·8°, 111·5°, 61·4°   16. £26,800, £37 200

34. 465·65, 501·05   35. (a) 16·7   (b) 15·8   (c) 15·5,
                              18·3

36. 14·2, 0·805, 5·67   37. 17·37, 16·95, 16, 18·61

38. Average quarterly relative deviations: 1·015, 1·07, 0·99, 0·925
    Seasonally adjusted figures: 228, 238, 246, 301, 309, 346, 415,
    382, 386, 325, 209, 217   39. 12·64, 10·55, 15·49, 24·42, 21·97,
    27·36

40. 49·027, 2284   41. 44·9, 8·2, 85·7%   42. 67·7, 90·8, 4·6%,
    6·0%

44. 46, 41·2   47. 97·4, 29·6 (last group 250–350)   49. 208·9

50. (a) 205,   (b) Co. of var. 9·5   (c) 197

54. 239·1, 241·6, 246·4, 249, 253·5, 253·9, 255·9, 256·7, 258·4, 258·6,
    261·3, 260·9   55. 103·7, 102·9, 101·1, 104·0, 106·4, 107·1,
    108·1, 109·5, 107·8, 99·3, 91·9, 85

57. Average seasonal variation: $Q_1 = 103$, $Q_2 = 94$, $Q_3 = 102$,
$Q_4 = 101$

Deseasonalised Figures

	Q_1	Q_2	Q_3	Q_4
1969	82·52	78·72	81·37	84·16
1970	82·52	84·04	88·24	92·08
1971	101·94	112·77	124·51	134·65
1972	139·81	141·49	144·12	130·69

58. Av. seasonal var. $Q_1 = 41$, $Q_2 = 77$, $Q_3 = 239$, $Q_4 = 43$
59. Average quarterly relative deviations: 0·998, 0·996, 1·010, 1·005
    Seasonally adjusted figures: 5308, 6033, 6018, 6099, 6075, 5683, 6219, 6448, 6845, 7784, 8127, 8801, 9280, 9640, 9148, 8942
63. 135·2, 133·7   65. 61, 140·7   66. 99·5   69. 135·85

70.

1971	97·7	100·7	101·2	100·4
1972	101·1	102·1	105·3	110·0
1973	116·1	125·3	137·5	151·6
1974	180·8	200·8	205·5	211·3

71. Median 34·62 yrs.   Quartile deviation 19·4 yrs.   72. 1 min.
73. 19·2, 13·96   76. (a) 6·3   (b) 12   81. 8·4
82. $y = -0·16x + 5·57$, 1982 = 2·69   83. 0·68   84. 0·79
85. 0·30   87. $-0·86$ 88. 0·89   90. 0·89

Index

319